Dilemmas and Difficulties
in the Management of Psychiatric Patients

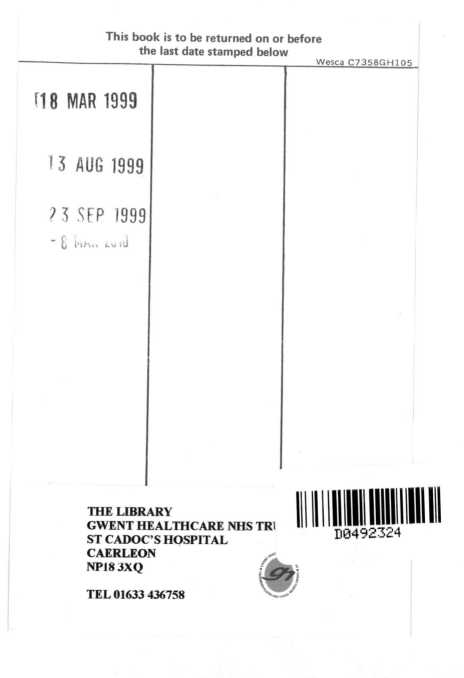

**This book is to be returned on or before
the last date stamped below**

Wesca C7358GH105

⌈18 MAR 1999

13 AUG 1999

23 SEP 1999

- 8 MAR 2010

Dilemmas and Difficulties in the Management of Psychiatric Patients

Edited by

KEITH HAWTON
Consultant Psychiatrist and Clinical Lecturer,
University Department of Psychiatry and
Warneford Hospital, Oxford

and

PHILIP COWEN
MRC Clinical Scientist,
Medical Research Council Clinical Pharmacology Unit,
Littlemore Hospital, Oxford

OXFORD   NEW YORK   TOKYO
OXFORD UNIVERSITY PRESS
1990

Oxford University Press, Walton Street, Oxford OX2 6DP
Oxford   New York   Toronto
Delhi   Bombay   Calcutta   Madras   Karachi
Petaling Jaya   Singapore   Hong Kong   Tokyo
Nairobi   Dar es Salaam   Cape Town
Melbourne   Auckland
and associated companies in
Berlin   Ibadan

Oxford is a trade mark of Oxford University Press

Published in the United States
by Oxford University Press, New York

British Library Cataloguing in Publication Data
Dilemmas and difficulties in the management of psychiatric
patients.
1. Psychiatric patients. Treatment
I. Hawton, Keith 1942–   II. Cowen
616.89
ISBN 0–19–261883–0
ISBN 0–19–261944–6 (pbk)

Library of Congress Cataloging in Publication Data
Dilemmas and difficulties in the management of psychiatric patients /
edited by Keith Hawton & Philip Cowen.
(Oxford medical publications)
Papers presented at a two-day meeting held at Green College,
Oxford, in April 1989.
1. Mental illness–Treatment–Congresses.   2. Psychotherapy–
–Congresses.   I. Hawton, Keith, 1942–   .   II. Cowen, Philip.
III. Series.
[DNLM: 1. Mental Disorders–therapy–congresses.   2. Professional
–Patient Relations–congresses.   WM 400 D576]
RC480.5.D54   1990   616.89'1—dc20   90–6857
ISBN 0–19–261883–0
ISBN 0–19–261944–6 (pbk.)

Set by Footnote Graphics, Warminster, Wiltshire

Printed in Great Britain by
Bookcraft (Bath) Ltd, Midsomer Norton, Avon

Preface

Failure of treatment is an inevitable part of medical practice: for many of the problems that confront us, current knowledge and therapy are sadly limited. In such areas management issues are often controversial and practitioners may face the dilemma of sticking to a treatment which is not proving beneficial or trying another approach with which they are less familiar.

Such dilemmas and difficulties are particularly common in psychiatry because of the complex nature of psychiatric disorder and the problems of investigating treatment and outcome in a scientific way. Nevertheless, we believe that in most cases sufficient information exists to allow the construction of rational management strategies.

The purpose of this book is to construct such management strategies by bringing together contributors who have a particular clinical and research interest in the problems posed. We have selected relatively common dilemmas and difficulties with which most psychiatrists and their colleagues will be familiar; for example, what to do when routine treatments fail? When should confidential information be revealed? We also have addressed issues that are less common but especially controversial; should ECT be used in schizophrenia? Is psychosurgery ever justified?

Each author was asked to review the area of controversy and difficulty and to discuss the scientific evidence and clinical experience relating to various management plans. The resulting chapters were reviewed in April 1989, at a two-day meeting of the contributors at Green College, Oxford, after which a final version was produced by each author. In this way we were able to draw on a wealth of clinical experience and scientific knowledge in the production of each chapter. Three extra chapters were obtained from contributors who were unable to attend the meeting in Oxford.

We asked each other to conclude their chapter with a series of brief clinical guidelines which we hope will assist other psychiatrists and their colleagues when they encounter such difficulties and dilemmas during their clinical practice. The contributions are, of course, statements of personal opinion but we believe that they are based on both scientific evidence and clinical experience and therefore reflect viewpoints which will be widely accepted.

We wish to thank Dista Products Ltd, Duphar Laboratories Ltd, E. Merck Ltd, and Upjohn Ltd whose generous support made the meeting of contributors at Green College possible. The meeting itself proved both educative and stimulating and we hope that the tone of informed and challenging debate has been preserved in the following chapters. Most of all, of course, we hope that the book will be of value to psychiatrists, both experienced and in training, as well as other professional colleagues in their management of these difficult problems.

Oxford   K.H.
October 1989   P.C.

Contents

Contributors

Thomas R. E. Barnes
Senior Lecturer in Psychiatry, Charing Cross and Westminster Medical School, Horton Hospital, Epsom, Surrey, UK

Aaron T. Beck
Professor of Psychiatry, Department of Psychiatry, University of Pennsylvania, Philadelphia, Pennsylvania, USA

Henrietta Bullard
Consultant Forensic Psychiatrist, Fair Mile Hospital, Wallingford, Berkshire, UK

Jose Catalan
Senior Lecturer in General Hospital Psychiatry, Charing Cross and Westminster Medical School, The Kobler Centre, St Stephen's Clinic, Chelsea, London, UK

Julie Chalmers
Consultant Psychiatrist, Elms Clinic, Horton General Hospital, Banbury, Oxfordshire, UK

Jonathan Chick
Consultant Psychiatrist, Royal Edinburgh Hospital; Senior Lecturer, Department of Psychiatry, University of Edinburgh, UK

David M. Clark
Lecturer in Psychology, Department of Psychiatry, University of Oxford; Fellow, University College, Oxford, UK

John Cobb
Consultant Psychiatrist, The Priory Hospital, Roehampton, London; Senior Clinical Tutor in Behavioural Psychotherapy and Honorary Senior Lecturer, Maudsley Hospital, London, UK

Philip J. Cowen
MRC Clinical Scientist, Medical Research Council Clinical Pharmacology Unit, Littlemore Hospital, Oxford, UK

Edna Foa
Professor, Department of Psychiatry, The Medical College of Pennsylvania, Philadelphia, Pennsylvania, USA

Michael Gelder
Professor of Psychiatry, Department of Psychiatry, University of Oxford, Warneford Hospital, Oxford, UK

Guy Goodwin
MRC Clinical Scientist, Medical Research Council Brain Metabolism
Unit, Royal Edinburgh Hospital, Edinburgh, UK

John R. Hamilton
Consultant Forensic Psychiatrist, Department of Health, London;
Honorary Senior Lecturer in Forensic Psychiatry, Institute of Psychiatry,
University of London, UK

Keith Hawton
Consultant Psychiatrist and Clinical Lecturer, University Department of
Psychiatry and Warneford Hospital, Oxford, UK

Mike Hobbs
Consultant Psychotherapist and Clinical Lecturer, Warneford Hospital,
Oxford, UK

Eve C. Johnstone
Professor of Psychiatry, Department of Psychiatry, Royal Edinburgh
Hospital, Edinburgh, UK

David Julier
Consultant Psychiatrist and Clinical Lecturer, Littlemore Hospital,
Oxford, UK

Desmond Kelly
Medical Director, Priory Hospital, Roehampton, London, UK

Gethin Morgan
Professor of Mental Health, Department of Mental Health, University of
Bristol, Bristol, UK

Eugene S. Paykel
Professor of Psychiatry, University of Cambridge, Addenbrooke's
Hospital, Cambridge, UK

Barbara Olasov Rothbaum
Assistant Professor, Department of Psychiatry, the Medical College of
Pennsylvania, Philadelphia, Pennsylvania, USA

Michael Sharpe
Lecturer in Psychiatry, Department of Psychiatry, University of Oxford,
Warneford Hospital, Oxford, UK

Pamela J. Taylor
Head of Medical Services, The Special Hospitals Service Authority,
London, UK and Honorary Senior Lecturer in Forensic Psychiatry,
Institute of Psychiatry, London, UK

Marjorie E. Weishaar
Clinical Assistant Professor, Psychiatry and Human Behaviour, Brown
University, Providence, Rhode Island, USA

1

Drug treatment of depression: what if tricyclics don't work?

GUY GOODWIN

Most patients who present to a general practitioner or a psychiatrist in the UK with a depressive illness will receive drug treatment. Most commonly this will be with a tricyclic antidepressant (TCA) and the majority of patients will recover. However a minority will not, and the responsible doctor will have to face the problem of what to do next. While the original diagnosis may have been wrong or the patient may not have taken the tablets, there is undoubtedly a significant proportion of depressed patients who simply do not get better when treated appropriately, if conventionally, with tricyclic drugs; a common estimate is 30 per cent (Quitkin 1985). This chapter will attempt to summarize the potential practical courses of action for such patients. Action is necessary because there is a high and perhaps poorly appreciated morbidity and mortality from chronic refractory depression that outcome studies, almost without exception, have revealed (Winokur and Morrison 1973; Keller *et al.* 1984; Lee and Murray 1988; Kilo *et al.* 1988).

Unfortunately, investigations of how to treat refractory patients which employ the hallowed ideal of a randomized double-blind clinical trial are remarkably few. Most practical suggestions result from open or anecdotal studies on small numbers of patients. There are good reasons why this is and will probably continue to be so: single centres yield too few patients, the numbers required for useful (i.e. statistically powerful) studies are large, and there is not sufficient interest in drug treatment as a special interest among general psychiatrists in this country although there may be elsewhere. It is therefore fortunate that besides the raw empiricism which is the strength of randomized clinical trials, there is a growing pharmaco-logical understanding of what antidepressant treatments do to specific neurotransmitter systems in the brain. Such knowledge can guide treat-ment strategies and may eventually assist our evaluation of their effective-ness. However, as a consequence, there will be a growing need for general psychiatrists to become more familiar with the pharmacology of the drugs they use. Finally, the phenomenology and clinical investigation of depressed patients assists, albeit only slightly, in prediction of treatment response (Joyce and Paykel 1989).

At present, the choice of what to do cannot be decided on a fully scientific basis and does not command a practical consensus. We enjoy a certain freedom to indulge our impulses for either dogma or disputation. The recommendations that follow are offered, and should be examined, with due scepticism.

Increase the dose

It is a central tenet of pharmacology that there exists a dose–response curve to describe the action of any drug. Until the dose has been increased to near the asymptote of this curve it makes no sense to say that a drug has not worked. We do not, of course, know what the crucial pharmacological action of antidepressant drugs is and how to measure it. But, in principle, we could determine the efficacy or probability of response at different fixed doses of antidepressant drugs. In practice, there have been very few studies to determine the dose–response curve even in part by this means. Two studies which have explicitly compared the responses to different fixed doses of imipramine and desipramine respectively (Simpson *et al.* 1976; Watt *et al.* 1972) showed a better response to the higher dose, 300 mg daily, than to the lower, 150 mg daily.

In general, the doses of tricyclic drugs actually employed in clinical practice appear to err seriously on the low side. Quitkin (1985), in reviewing the evidence for this, cited first the published data from the treatment regimes of patients who were designated refractory; secondly, the recommended prescribing levels; and thirdly the comparison treatment dose for standard drugs in treatment trials. For TCAs, 150 mg daily of imipramine or its equivalent is regarded as adequate treatment. This is reinforced by recommended doses in, for example, the British National Formulary and from manufacturers. By contrast, in the United States where 'psychopharmacology' is an emergent sub-specialty within psychiatry, the recommendation of higher doses is becoming commonplace. Quitkin (1985), for example, reports that 30–80 per cent of patients deemed refractory to the standard dose of tricyclic will respond when treated with a higher dose. Bridges (1983) made the same point from a British perspective.

In practice, no patient should be deemed refractory to a tricyclic unless at least 150 mg (for amitriptyline, the same or equivalent dose for other compounds) has been employed for an adequate length of time (see below). The dose can and probably should be raised towards 300 mg daily, with appropriate allowance for body mass. Inability to tolerate side-effects limits this strategy. A plasma level, if low, may reinforce the decision to push the dose up further. However, it may be noticed that plasma levels of tricyclics, like oral doses, have not been investigated adequately in relation to response (reviewed by Preskorn and Mac 1984). Accordingly, plasma levels offer no useful guide to defining adequacy of treatment. The idea of a 'therapeutic level' sounds reassuring but has not been determined any better than a 'therapeutic dose'.

If side-effects allow, the dose of tricyclic can even be increased beyond 300 mg, with ECG monitoring. American practice is to accept gradual increases in dosage of 50 mg up to 500 mg daily while the pulse rate remains below 100/min and the PR interval and QRS and T waves remain within normal limits. Daily doses of 500 mg of desipramine may be safe given these precautions, but the published data is not sufficient to prove that this manoeuvre is actually the best of the alternatives for treating the patient. Furthermore the time required slowly to increase the dose into this range while waiting to observe a measurable treatment response may be too long to make this option attractive except in unusual circumstances.

In summary, an increase in tricyclic dose as high as side-effects permit (to a maximum of 300 mg daily) is required before the drug can be declared ineffective. Furthermore, a duration of treatment of at least four and preferably six weeks is required before any clinical judgement of non-response should be made (Quitkin *et al.* 1984). A further increase in dose offers a treatment option to compete with alternatives described below. It should be kept in mind that 90 per cent response rates have been reported for monotherapy with desipramine in non-psychotic patients displaying prominent melancholic (endogenous) symptoms (Nelson *et al.* 1982).

Try another drug

Whatever the strictures of the preceding paragraph, it appears likely that once a tricyclic has been prescribed at a dose of 150 mg for four weeks, a trial of another antidepressant is commonly what doctors actually propose. What then is the evidence that another class of treatment will be effective where a previous one has failed?

A further re-uptake inhibitor

Most of the alternative antidepressant drugs that inhibit monoamine re-uptake (e.g. nortriptyline, protriptyline, butriptyline, doxepine, dothiepin, trimipramine) closely resemble the prototype drugs and may only differ from them in having fewer anticholinergic actions and hence associated side-effects. Thus, they may be better tolerated, but given their otherwise similar pharmacology, it would be surprising if they were more effective than the prototype drugs. Other drugs such as iprindole, trazodone, lofepramine and mianserin have less well characterized modes of action and accordingly may have some claim to be different from re-uptake inhibitors; do they ever provide a superior alternative in refractory patients? No evidence from controlled trials has ever shown that they do. Switching to them, except to avoid particularly severe cholinergic side-effects, is probably a waste of time. However, switching *from* a compound of uncertain absolute efficacy to a standard tricyclic such as imipramine, amitriptyline, or desipramine appears eminently sensible.

Selective inhibitors of monoamine re-uptake

The selective re-uptake inhibitors require separate consideration. Desipramine is a metabolite of imipramine and is itself moderately selective for noradrenergic neurotransmission. Accordingly, more selective drugs for noradrenaline (NA) re-uptake, such as maprotiline, offer little advantage over desipramine, the clinical pharmacology of which has been characterized as well as that of any other antidepressant. Furthermore, the marked dose-related side-effect of postural hypotension complicates all noradrenergic treatment strategies.

Drugs selective for the 5-hydroxytryptamine (5-HT) transporter are still relatively new. Clomipramine and zimeldine (no longer available) have been shown to be effective antidepressants in placebo-controlled trials but both have metabolites that are pharmacologically active in blocking NA re-uptake. The highly potent and selective compounds fluvoxamine, fluoxetine, and paroxetine offer a new order of selectively and have been shown to be effective antidepressants in placebo-controlled trials. It had been hoped that they would offer an additional dimension to drug treatment because some patients when depressed appear to have reduced 5-HT turnover. Unfortunately, the preliminary evidence from well controlled trials comparing citalopram and, more recently paroxetine with clomipramine suggest that the highly selective compounds are *less* effective than clomipramine in the treatment of major depressive disorder (Danish University Antidepressant Group 1986, 1990). In addition, while it may still be too early to be emphatic, the only controlled trial that has addressed the efficacy of a selective 5-HT re-uptake inhibitor in 'refractory patients' was entirely negative (Nolen *et al.* 1988*a*). Under identical circumstances, the same group demonstrated that the change to a monoamine oxidase inhibitor (MAOI) was effective as a treatment strategy. The negative findings with both fluvoxamine and a selective NA re-uptake inhibitor, oxaprotiline suggest that selective re-uptake inhibitors offer no advantages in terms of efficacy over non-selective drugs.

Selective inhibitors of dopamine re-uptake have not been properly evaluated and the withdrawal of nomifensine removed the only drug displaying this property to any significant degree among current antidepressant drugs. In conclusion, the evidence does not support the strategy of trying a selective re-uptake inhibitor after a standard tricyclic has been given an adequate trial.

Antioxidants, light, and cognitive therapy

The foregoing conclusion applies, only more so, to the use, as alternatives, of special interest treatments such as methylene blue, vitamin C, light, and cognitive therapy. It seems logical that we strive to use more effectively the treatments that we are confident can work in refractory patients before taking up those for which so much cannot be assumed.

Something completely different: ECT or MAOIs

A change from a tricyclic to either electroconvulsive therapy (ECT) or a monoamine oxidase inhibitor (MAOI) really does represent a shift to a different modality of treatment. Historically, ECT and MAOIs were discovered independently of each other and of the TCAs. Pharmacologically both ECT and MAOIs share important similarities of action with TCAs but there are also important differences because ECT and MOAIs act prominently on dopamine systems (Green *et al.* 1986). Clinically, a change to either ECT or an MAOI is a defensible first move in treating refractory depressive illness.

Electroconvulsive therapy

In the clinical trial of the treatment of depressive illness organized by the Medical Research Council (MRC) (1965), a 50 per cent response rate to ECT was described in patients failing to respond to drug treatment. Recent trials of ECT have again underlined its efficacy and the Leicestershire trial (Brandon *et al.* 1984) is particularly noteworthy because entry to the study and randomization to ECT or sham treatment occurred after routine referral for ECT. It therefore reflected actual current practice and will have included many patients refractory to drug treatment.

Monoamine oxidase inhibitors

MAOIs emerged poorly from the MRC trial (1965), probably because the dose of phenelzine employed was too low. However, their use has continued to be widespread and there is important evidence that MAOIs can be effective in treatment of tricyclic-resistant patients. Thus, in a rare example of a randomized trial of treatment in refractory patients, tranylcypromine was given to patients refractory to tricyclics (150 mg for four weeks), and gave a response rate of 50 per cent (Nolen *et al.* 1988*b*); this study was performed in the Netherlands where ECT is not a practical treatment option for political reasons (Nolen *et al.* 1985).

ECT is most effective in retarded and deluded patients, while MAOIs appear most efficacious where anxiety symptoms are particularly evident. However, where tranylcypromine was effective in refractory patients to whom ECT was not available, it worked well in those with endogenous symptoms (Nolen *et al.* 1988*b*). Since ECT is difficult to use routinely in outpatients, MAOIs may be the best choice in that setting. Electroconvulsive therapy may be the better choice for in-patients. The use of MAOIs is described at greater length in Chapter 2. The Royal College of Psychiatrists (1989) has recently published revised guidelines for the use of ECT, which conclude that bilateral ECT appears to be a more effective option than unilateral application.

Add another drug

Combination treatment after the failure of a tricyclic drug has attracted a good deal of interest in recent years. The original combination approach added an MAOI to the tricyclic. This appears to have developed as an empirical solution to the problems of refractoriness and it carries a reputation for inherent danger. These dangers appear greatest in the simultaneous or sequential use of 5-HT selective re-uptake inhibitors (clomipramine or the newer drugs) and an MAOI; the combination is also dangerous in overdose (see Marley and Wozniak 1983). However, addition of an MAOI to a *non-selective* tricyclic regimen, or, preferably, their joint administration after withdrawal of tricyclic treatment, is safe (Pare 1985). What is more uncertain is the effectiveness in resistant patients, and whether the combination is superior to one or other drug used alone. A non-selective re-uptake inhibitor taken together with an MAOI tends to reduce but not exclude the risk of a tyramine related 'cheese reaction' (Pare 1985).

The neuropharmacology of tricyclic augmentation

Our increasing understanding of the neuropharmacological effects exerted by antidepressant drugs now permits a proper account of how combination treatments may work or be devised. All antidepressant treatments appear to increase transmission across monoamine synapses. This should not be understood as a simple reversal of what is amiss in depression. However this pharmacological intervention allows recovery to take place with a higher probability than placebo. It follows that failure to recover may reflect inadequate pharmacology or an unspecifiable refractoriness of the depressed state. Our ability to enhance monoamine neurotransmission is illustrated for the neurotransmitter 5-hydroxytryptamine (5-HT) in Figure 1. It summarizes the stages believed to be involved in the synthesis, release, and action of the neurotransmitter. It also indicates the loci at which drugs are believed to work.

1.　The presynaptic *availability* of a neurotransmitter may be influenced by the supply of precursor, the rate of degradation by monoamine oxidase (MAO), and the rate of re-uptake of transmitter. Drugs acting at these loci would be expected to be synergistic and, indeed, appear to be; the evidence for MAOIs has already been summarized.

2.　To act, a neurotransmitter must be *released* efficiently. Drugs that increase the release of neurotransmitters or inhibit autoreceptor inhibition of release could increase the efficacy of TCAs or be effective in their own right.

3.　The postsynaptic actions of a neurotransmitter may be modulated by receptor blockers or modification of second messenger systems (an important action of lithium). Such actions may again serve to boost transmission across the synapse.

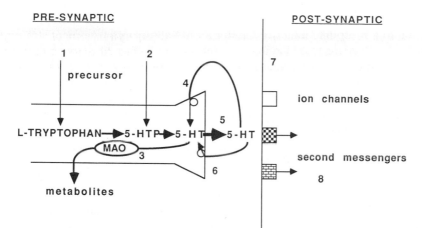

Fig. 1. *Diagram to illustrate mechanisms involved in transmission across a 5-hydroxytryptamine synapse.* Numbers indicate loci for drug action. 1 and 2: Uptake of precursor can be supplemented by L-tryptophan or 5-hydroxytryptophan (5-HTP). 3: Degradation of transmitter by monoamine oxidase (MAO) can be prevented by MAOIs. 4: Re-uptake of neurotransmitter can be blocked by tricyclic and other drugs of varying potency and selectivity. 5: Release of 5-hydroxytryptamine (5-HT) can be facilitated by amphetamine-like drugs. 6: Autoreceptor inhibition of transmitter release can be prevented by receptor antagonists (which will also bind post-synaptic receptors, 7). 8: Second-messenger systems are influenced by lithium.

There is a certain amount of clinical evidence that supports the validity of the pharmacological theory, and this is reviewed in the next sections. In addition, two synergistic treatments may produce a desired therapeutic effect at doses that avoid side-effects characteristic of higher doses of either drug used alone. However, theoretical considerations should not be over-stated in the absence of adequate treatment trials allowing comparison between treatment options.

Neurotransmitter precursors

Oral tryptophan, either alone or in combination with amitriptyline, is superior to placebo in the treatment of out-patient depression (Thomson *et al.* 1982). Tryptophan is the precursor for 5-HT and following oral administration at doses of about 3 g increased levels of 5-HT and its metabolites can be detected in cerebro-spinal fluid (Gillman *et al.* 1981). Tryptophan is converted to 5-hydroxytryptophan (5-HTP) in the synthesis of 5-HT. Both precursors have been used as antidepressants, but neither now appears to offer any advantage as an alternative to tricyclic drugs in refractory patients. Specifically, 5-HTP, up to 200 mg daily, (together with the peripheral decarboxylase inhibitor, carbidopa) was ineffective when given alone to such patients (Nolen *et al*. 1988*b*). However, addition of tryptophan or 5-HTP to a re-uptake inhibitor has apparent validity as a treatment strategy. It has been advocated by numerous authors in the last 20 years on the basis of open trials or clinical experience (reviewed by Nolen *et al*. 1985). Tryptophan or 5-HTP in combination with amitriptyline or clomipramine appears safe. With the more potent modern 5-HT re-uptake inhibitors, however, toxic effects with tryptophan or 5-HTP have been reported and care should be exercised if clinicians decide to use these drugs in combination. Furthermore, it is possible that adding precursor to the highly selective 5-HT re-uptake inhibitors does not improve their performance in any case (Walinder *et al*. 1980, 1981).

While a strategy of precursor supplementation for 5-HT systems has been quite widely adopted, noradrenergic or dopaminergic precursors have been less explored. There is some scientific rationale for this selection because tryptophan availability is more clearly rate-limiting for transmitter synthesis than is the availability of other monoamine precursors (Fernstrom 1983). Nevertheless addition of tyrosine or l-DOPA would appear an entirely logical strategy for use with noradrenergic or non-selective re-uptake inhibitors. Preliminary experience appears to have been negative but not conclusive (reviewed briefly by Schmauss 1983).

Finally, it has sometimes been proposed that co-factors of neurotransmitter synthesis rather than precursors may be deficient. S-adenosyl-l-methionine (SAM), a methyl donor, has been advocated as an antidepressant (Agnoli *et al*. 1976) and, speculatively, may play some role of this sort. Perhaps similarly, folate, which is also involved in transmethylation, (at an appropriately *low* dose of 0.2 mg) has been reported to facilitate the control of affective symptoms in patients treated prophylactically with lithium (Coppen *et al*. 1986).

Releasing agents

Drugs that release neurotransmitters include amphetamine, reserpine (both non-selective for monoamines), and fenfluramine (mainly 5-HT). Amphetamine was widely prescribed for depressive illness by an earlier generation of psychiatrists but its value now in refractory patients is uncertain. Wager and Klein (1988) believe amphetamine merits wider use in combination with, or as an alternative to, tricyclics. Speculatively, a similar use could be found for fenfluramine. It should be noticed that tranylcypromine (an MAOI) has amphetamine-like properties that may contribute to its proven efficacy (p. 5).

There are also reports from 20 years ago of reserpine producing marked antidepressant action when used to augment the treatment response to a tricyclic. However, recent open studies have not confirmed this finding (literature reviewed by Price *et al*. 1987), and this strategy cannot be recommended.

Lithium

Lithium has well-known uses both as an anti-manic agent and in prophylaxis of bipolar illness. Its use as an adjunct to other drugs in depressive illness is relatively recent. It is interesting in that it grew out of basic investigation of the mechanism of action of antidepressant drugs in animals. The specific claim by de Montigny *et al.* (1983) was of a very rapid therapeutic response (within 48 hours) following the introduction of lithium in patients refractory to re-uptake inhibitors. This was not fully confirmed under placebo-controlled conditions by Heninger *et al.* (1983), but anecdotally, certainly occurs in some patients. Price *et al.* (1985) also described augmentation of response to MAOIs by lithium. Katona (1988) may be consulted for a recent review of lithium's clinical use in refractory depression.

Lithium augmentation may be an alternative to ECT even in deluded patients according to Pai *et al.* (1986); further, Price *et al.* (1983) claim augmentation by lithium has the therapeutic efficacy of tricyclic/neuroleptic treatment in psychotic depression. This may be important in practice when it is difficult to obtain consent to ECT, but patients remain compliant with drug treatment.

A particular option that should be noted is to add tryptophan *and* lithium to clomipramine as the ultimate combination treatment. This has a reputation for unusual efficacy (Hale *et al.* 1987) and appears to be safe. However, addition of lithium to the new high potency 5-HT re-uptake inhibitors requires caution since seizures have been reported on such combinations (Committee on Safety of Medicines 1989). Combination treatment with an MAOI, tryptophan, and lithium has also been advocated from open studies of refractory patients (see Katona 1988). However, there are dangers in the latter combination; Brennan *et al.* (1988) describe fatal malignant hyperpyrexia with phenelzine, lithium, and tryptophan.

A full discussion of the likely mechanism of the effect of lithium is beyond the scope of this chapter (see Wood and Goodwin 1987). However, Price *et al.* (1989) have recently suggested that the pharmacological effect of lithium upon 5-HT function wears off in patients after several weeks. If facilitation of 5-HT neurotransmission is the important mechanism, the value of maintaining patients on lithium augmentation following their recovery may be questionable unless it is wanted for long-term prophylaxis.

Neuroleptics and other receptor antagonists

Addition of neuroleptics to tricyclic treatment is regarded as an important alternative to ECT in psychotically depressed patients (but see Paykel and Hale 1986). Addition of neuroleptic may simply reduce psychotic symptoms, as occurs in psychotic illness irrespective of diagnosis (Johnstone *et al.* 1988). High dose dopamine antagonists are likely to block postsynaptic mechanisms. However, lower doses of neuroleptic may enhance dopamine function by inhibiting autoreceptors. This is believed to underlie the anti-depressant action of flupenthixol. The existence of autoreceptors in monoamine neurones means that antagonists will simultaneously promote as well as block synaptic transmission in these systems. Thus alpha$_2$-adrenoceptor antagonists and 5-HT1-like receptor antagonists will, when available, have interesting potential for modifying NA or 5-HT function respectively, alone or as an adjunct to re-uptake inhibitors. Against this possibility, yohimbine, a poorly selective alpha$_2$-adrenoceptor antagonist, failed to augment the clinical effects of desipramine despite enhancing some of the adaptive receptor changes produced by desipramine in animals (Charney *et al.* 1986).

Hormones

Goodwin *et al.* (1982) suggested a trial of tri-iodo-thyronine as an augmenting treatment with a tricyclic in resistant patients. However, neither the original data nor subsequent unsuccessful efforts to replicate the findings (Gitlin *et al* 1987 and research reports cited by Wager and Klein 1988) allow any confidence in this approach. Further work is needed before it can be seriously recommended. A particular worry is that in older patients it may be hazardous. Glucocorticoids are also mood elevating in some individuals at high doses and have been claimed to be effective in treating refractory patients (Kurland 1965). Unfortunately this report is too flimsy to interpret and glucocorticoids have not been seriously explored as anti-depressants or as adjuncts to other treatments.

　　Finally, it is an old idea that depression after the menopause might be relieved by treatment with oestrogen. Some modest effects in quite severely depressed pre-and post-menopausal women have been reported by Klaiber *et al.* (1979) for treatment with high doses of oestrogens. There appears to have been little interest recently in pursuing this approach in refractory patients.

Psychological treatments

Refractory illnesses tend to be long illnesses and, it appears, vice versa. Despair of ever recovering, losses of relationships, social role, and almost invariably, employment are common problems. Recovery from chronic depression, at least in the community, may require patients to respond emotionally to 'fresh start' or other life events (Brown *et al.* 1988). Further, the failures of drug treatment are likely to be seen by the patient both as incompetence by the responsible doctor and as reason for further despair. For all these reasons, management of refractory patients is assisted by a planned psychological strategy that both staff and patients understand. This may simply require the provision of a supportive milieu in day- or in-patient setting, but at times a focused psychotherapy dealing with grief, marital difficulties, or family dynamics seems helpful. Particular claims are also sometimes made for cognitive therapy as an additional component of the treatment package in refractory patients. However, a balanced discussion of this area soon subsumes most of current psychiatric practice. In any case, opinion remains more often expressed than evaluated.

Psychosurgery

Chronic refractory depressive illness is potentially an indication for psychosurgery. The patients who come through all the above treatment options without response will have been in a severely depressed state, probably as in-patients, for over a year. Apart from the subjective suffering that this implies (Gray (1983) gives a graphic personal account) and the many potentially irreversible losses that are incurred by being depressed, there is also a high risk of suicide. Benefits are claimed for psychosurgery in severe states of this sort with a recovery rate of 68 per cent quoted by Goktepe *et al.* (1975). It is sometimes regarded as difficult to defend psychosurgery; it is equally difficult to defend therapeutic nihilism when one's conventional options have run out (see Chapter 18 for a fuller discussion of psychosurgery).

Clinical guidelines

If treatment with a tricyclic drug fails to reduce depressive symptoms the following steps are recommended.

1. The dose should be increased to the maximum tolerated or 300 mg daily (equivalent of desipramine); four to six weeks should be allowed for assessment of treatment response.

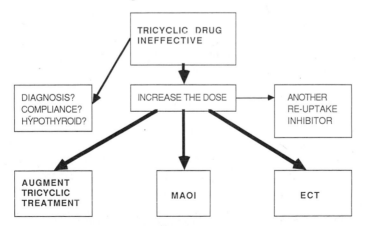

Fig. 2. Decision tree after finding a tricyclic drug ineffective. If a dose increase either fails or is impossible, there is a clinical choice to make between augmenting the tricyclic treatment, changing to an MAOI or employing ECT. There is not a single route through these options which is why they are shown as equals.

2.   If patients are clearly refractory to adequate tricyclic treatment, the diagnosis and treatment compliance should be reviewed.

3.   There is no universally applicable next step. All medical treatment should be decided by a proper weighing of costs and benefits to individual patients. In many ways we are clearer about costs such as drug side-effects, risk of over-dose, dietary modification etc., than we are about the relative benefits of different treatments. However, the options are summarized in Fig. 2. Lithium augmentation appears to combine effectiveness, speed of response, and simplicity. However, there is a need for its further empirical investigation. The use of MAOIs or ECT carries the best proven evidence of likely benefit.

4.   The use of more adventurous combination treatments may be worthwhile but this carries the risk of adverse reactions, and probably requires supervision by a psychiatrist with a special interest and experience in psychopharmacology. Recognition of the need for a greater emphasis on pharmacological therapeutics in general psychiatric training in the UK is awaited.

5.   As a patient's illness becomes increasingly prolonged, the need for empirical psychosocial intervention will become increasingly necessary. It should be provided in a form as free as possible from psychological dogma.

References

Agnoli, A., Andreoli, V., Casacchia, M., and Cerbo, R. (1976). Effects of S-adenosyl-L-methionine (SAMe) upon depressive symptoms. *Journal of Psychiatry Research*, **13**, 43–7.

Brandon, S., Cowley, P., McDonald, C., Neville, P., Palmer, R., and Wellstood-Eason, S. (1984). Electroconvulsive therapy: results in depressive illness from the Leicestershire trial. *British Medical Journal*, **228**, 22–5.

Brennan, D., MacManus, M., Howe, J., and McLoughlin, J. (1988). 'Neuroleptic malignant syndrome' without neuroleptics. *British Journal of Psychiatry*, **152**, 578–9.

Bridges, P. K. (1983). ' . . . and a small dose of an antidepressant might help'. *British Journal of Psychiatry*, **142**, 626–8.

Brown, G. W., Adler, Z., and Bifulco, A. (1988). Life events, difficulties and recovery from chronic depression. *British Journal of Psychiatry*, **152**, 487–98.

Charney, D. S., Price, L.H., and Heninger, G.R. (1986). Desipramine-yohimbine combination treatment of refractory depression: implications for the β-adrenergic receptor hypothesis of antidepressant action. *Archives of General Psychiatry*, **43**, 1155–61.

Committee on Safety of Medicines (1989). *Current Problems*, **26**, 3.

Coppen, A., Chaudry, S., and Swade, C. (1986). Folic acid enhances lithium prophylaxis. *Journal of Affective Disorders*, **10**, 9–13.

Danish University Antidepressant Group (1986). Citalopram: clinical effect profile in comparison with clomipramine. A controlled multicentre study. *Psychopharmacology*, **90**, 131–8.

Danish University Antidepressant Group (1990). Paroxetine: a selective serotonin re-uptake inhibitor showing better tolerance, but weaker antidepressant effect than clomipramine in a controlled multicentre study. *Journal of Affective Disorders* (in press).

de Montigny, C., Cournoyer, G., Morisette, R., Langlois, R., and Caille, G. (1983). Lithium carbonate addition in tricyclic antidepressant–resistant unipolar depression. *Archives of General Psychiatry*, **40**, 1327–34.

Fernstrom, J. D. (1983). Role of precursor availability in control of monoamine biosynthesis in brain. *Physiological Reviews*, **63**, 484–546.

Gillman, P. J., Bartlett, J. R., Bridges, P. K., Hunt, A., Patel, A. J., Kantamaneni, B. D., and Curzon, G. (1981). Indolic substances in plasma, cerebrospinal fluid and frontal cortex of human subjects infused with saline or tryptophan. *Journal of Neurochemistry*, **37**, 410–17.

Gitlin, M. J., Weiner, H., Fairbanks, L., Hershman, J. M., and Friedfeld, N. (1987). Failure of T_3 to potentiate tricyclic antidepressant response. *Journal of Affective Disorders*, **13**, 267–72.

Goktepe, E. O., Young, L. B., and Bridges, P. K. (1975). A further review of the results of stereotactic subcaudate tractotomy. *British Journal of Psychiatry*, **126**, 270–80.

Goodwin, F. K., Prange, A. J., Post, R. M., Muscettola, G., and Lipton, M. A. (1982). Potentiation of antidepressant effects by L-triiodothyronine in tricyclic nonresponders. *American Journal of Psychiatry*, **139**, 34–8.

Gray, F. G. (1983). Severe depression: a patient's thoughts. *British Journal of Psychiatry*, **143**, 319–22.

Green, A. R., Heal, D. J., and Goodwin, G. M. (1986). The effects of electro-convulsive therapy and antidepressant drugs on monoamine receptors in rodent brain—similarities and differences. In *Antidepressants and receptor function*(ed. R. Porter, G. Bock and S. Clark) pp. 246–261. CIBA Foundation Symposium 123. Wiley, Chichester.

Hale, A. S., Procter, A. W., and Bridges, P. K. (1987). Clomipramine, tryptophan and lithium in combination for resistant endogenous depression: seven case studies. *British Journal of Psychiatry*, **151**, 213–17.

Heninger, G. R., Charney, D. S., and Sternberg, D. E. (1983). Lithium carbonate augmentation of antidepressant treatment. *Archives of General Psychiatry*, **40**, 1335–9.

Johnstone, E. C., Crow, T. J., Frith, C. D., and Owens D. G. C. (1988). The Northwick Park 'functional' psychosis study: diagnosis and treatment response. *Lancet*, **ii**, 119–24.

Joyce, P. R. and Paykel, E. S. (1989). Predictors of drug response in depression. *Archives of General Psychiatry*, **46**, 89–99.

Katona, C. L. E. (1988). Lithium augmentation in refractory depression. *Psychiatric Developments*, **2**, 153–71.

Keller, M. B., Klerman, G. L., Lavori, P. W., Coryell, W., Endicott, J., and Taylor, J. (1984). Long-term episodes of major depression: clinical and public health significance. *Journal of the American Medical Association*, **252**, 788–92.

Kiloh, L. G., Andrews, G. and Neilson, M. (1988). The long-term outcome of depressive illness. *British Journal of Psychiatry*, **153**, 752–7.

Klaiber, E. L., Broverman, D. M., Vogel, W., and Kobayashi, Y. (1979).

Estrogen therapy for severe persistant depressions in women. *Archives of General Psychiatry*, **36**, 550–5.

Kurland, H. D. (1965). Physiologic treatment of depressive reactions: a pilot study. *American Journal of Psychiatry*, **122**, 457–8.

Lee, A. S. and Murray, R. M. (1988). The long-term outcome of Maudsley depressives. *British Journal of Psychiatry*, **153**, 741–51.

Marley, E. and Wozniak, K. M. (1983). Clinical and experimental aspects of interactions between amine oxidase inhibitors and amine re-uptake inhibitors. *Psychological Medicine*, **13**, 735–49.

Medical Research Council (1965). Clinical trial of the treatment of depressive illness. *British Medical Journal*, **1**, 881–6.

Nelson, J. C., Jatlow, P., Quinlin, D. M., and Bowers, M. B. (1982). Desipramine plasma concentrations and antidepressant response. *Archives of General Psychiatry*, **39**, 1419–22.

Nolen, W. A., van de Putte, J. J., Dijken, W. A., and Kamp, J. S. (1985). L-5HTP in depression resistant to re-uptake inhibitors; an open comparative study with tranylcypromine. *British Journal of Psychiatry*, **147**, 16–22.

Nolen, W. A., van de Putte, J. J., Dijken, W. A., Kamp, J. S., Blansjaar, B. A., Kramer, H. J., and Haffmans, J. (1988*a*). Treatment strategy in depression. I. Non-tricyclic and selective reuptake inhibitors in resistant depression: a double-blind partial crossover study on the effects of oxaprotiline and fluvoxamine. *Acta Psychiatrica Scandinavica*, **78**, 668–75.

Nolen, W. A., van de Putte, J. J., Dijken, W. A., Kamp, J. S., Blansjaar, B. A., Kramer, H. J., and Haffmans, J. (1988*b*). Treatment strategy in depression. II. MAO inhibitors in depression resistant to cyclic antidepressants: two controlled crossover studies with tranylcypromine versus L-5-hydroxytryptophan and nomifensine. *Acta Psychiatrica Scandinavica*, **78**, 676–83.

Pai, M., White, A. C., and Deane, A. G. (1986). Lithium augmentation in the treatment of delusional depression. *British Journal of Psychiatry*, **148**, 736–8.

Pare, C. M. B. (1985). The present status of monoamine oxidase inhibitors. *British Journal of Psychiatry*, **146**, 576–84.

Paykel, E. S. and Hale, A. S. (1986). Recent advances in the treatment of depression. In *The biology of depression* (ed. J.F.W. Deakin), pp. 153–73. The Royal College of Psychiatrists, Gaskell, London.

Preskorn, S. H. and Mac, D. S. (1984). The implications of concentration/response studies of tricyclic antidepressants for psychiatric research and practice. *Psychiatric Developments*, **3**, 201–22.

Price, L. H., Conwell, Y., and Nelson, J. C. (1983). Lithium augmentation of combined neuroleptic tricyclic treatment in delusion depression. *American Journal of Psychiatry*, **140**, 318–22.

Price, L. H., Charney, D. S., and Heninger, G. R. (1985). Efficacy of lithium-tranylcypromine treatment in refractory depression. *American Journal of Psychiatry*, **142**, 619–23.

Price, L. H., Charney, D. S., and Heninger, G. R. (1987). Reserpine augmentation of desipramine in refractory depression: clinical and neurobiological effects. *Psychopharmacology*, **92**, 431–7.

Price, L. H., Charney, D. S., Delgado, P. L., and Heninger, G. R. (1989). Lithium treatment and serotonergic function. *Archives of General Psychiatry*, **46**, 13–19.

Quitkin, F. M. (1985). The importance of dosage in prescribing antidepressants. *British Journal of Psychiatry*, **147**, 593–7.

Quitkin, F. M., Rabkin, J. G., Ross, D., and McGrath, P. J. (1984). Duration of antidepressant treatment. *Archives of General Psychiatry*, **41**, 238–45.

Royal College of Psychiatrists: ECT subcommittee of the Research Committee (1989). *The practical administration of electroconvulsive therapy (ECT)*. Gaskell, London.

Schmauss, M. (1983). Miscellaneous new developments. In: *Psychopharmacology 1, Part 2: Clinical psychopharmacology* (eds. H. Hippius and G. Winokur) pp. 178–87. Excerpta Medica, Amsterdam.

Simpson, G. M., Lees, J. H., Cuculic, Z., and Kellner, R. (1976). The dosages of imipramine in hospitalized endogenous and neurotic depressives. *Archives of General Psychiatry*, **33**, 1093–102.

Thomson, J., Rankin, H., and Ashcroft, G. W. (1982). The treatment of depression in general practice: a comparison of L-tryptophan, amitriptyline, and a combination of L-tryptophan and amitriptyline with placebo. *Psychological Medicine*, **12**, 741–51.

Wager, S. G. and Klein, D. F. (1988). Drug treatment strategies for treatment-resistant depression. *Psychopharmacology Bulletin*, **24**, 69–74.

Walinder, J., Carlsson, A., Persson, R., and Wallin, L. (1980). Potentiation of the effect of antidepressant drugs by tryptophan. *Acta Psychiatrica Scandinavica* Suppl. 280, **61**, 243–9.

Walinder, J., Carlsson, A., and Persson, R. (1981). 5-HT reuptake inhibitors plus tryptophan in endogenous depression. *Acta Psychiatrica Scandinavica* Suppl. 290, **63**, 179–90.

Watt, D. C., Crammer, J. L., and Elkes, A. (1972). Metabolism, anticholinergic effects and therapeutic effects on outcome of desmethylimipramine in depressive illness. *Psychological Medicine*, **2**, 397–405.

Winokur, G. and Morrison, J. (1973). The Iowa 500: follow-up of 225 depressives. *British Journal of Psychiatry*, **123**, 543–8.

Wood, A. J. and Goodwin, G. M. (1987). A review of the biochemical and neuropharmacological actions of lithium. *Psychological Medicine*, **17**, 579–600.

2

Monoamine oxidase inhibitors: when should they be used?

EUGENE S. PAYKEL

Introduction

The monoamine oxidase (MAOIs) were introduced in the 1950s, a little earlier than the tricyclic antidepressants. In spite of a promising start their place remains less secure, with considerable differences between countries. In the UK use has remained comparatively high, but only among psychiatrists, rather than general practitioners. A number of problems have constrained prescribing, most notably fears about side-effects. Defining their specific place requires consideration of side-effects, overall efficacy, responsive clinical subgroups, biochemical predictors, and the place of alternative treatments.

Side effects

Cheese reaction

It was recognition of the cheese reaction (Blackwell 1963) which led to a rapid diminution in use. Prior to this there had been reports of hypertensive attacks, but the mechanism had been obscure. In fact the common side-effects are not the food and drug interactions, but dose-related hypotension, sleep disturbance, and ankle oedema. However it is the interactions which are dangerous, principally those with food and drugs containing indirectly acting pressor amines, particularly tyramine. Interactions with tricyclics, opiates, and insulin produce lesser problems.

There are no accurate data on the frequency of the cheese reaction. The introduction of tranylcypromine in the early 1960s coincided with its recognition, and the effect appears more common with this drug. Clinical experience suggests that, with sensible management and an intelligent patient, it is not a major problem. Nevertheless, the list of substances to be avoided is bewildering to anyone not trained as a pharmacologist and food biochemist, and there can be much uncertainty when faced with new foods, in everyday social situations. Batches of the same food may vary in tryamine content.

Efficacy

At around the time when the cheese reaction was recognized, doubts also started to grow as to the efficacy of MAO inhibitors. Several large collaborative studies on in-patients found them no better than placebo while the same studies, or parallel ones by the same research groups, found tricyclics effective (Greenblatt *et al*. 1964; Medical Research Council 1965; Overall *et al*. 1962).

Studies against placebo

Table 1 summarizes the principal controlled trials against placebo, in depressed samples, of the most commonly used MAO inhibitors, phenelzine, isocarboxazid, and tranylcypromine. Iproniazid is omitted in view of its hepatotoxicity.

The overall evidence is moderately good, but not as good as for the tricyclics, where, in the review by Morris and Beck (1974) of the drugs available in the USA, 61 out of 93 studies found the drug superior to placebo. For phenelzine, ten studies have shown the drug clearly superior to placebo, three were doubtfully positive, and four were negative. For isocarboxazid, three trials have been positive, three negative; one of the positive studies (Davidson *et al*. 1988) was a large collaborative trial in which the three centres also published separately. Tranylcypromine, believed by many to be the most effective MAO inhibitor, has also been the least rigorously evaluated: two studies have been positive, one doubtfully positive, and one negative.

Added to this is evidence of efficacy in phobic patients in whom iproniazid has been found superior to placebo (Lipsedge *et al*. 1973) as has phenelzine (Sheehan *et al*. 1980; Solyom *et al*. 1973; Tyrer *et al*. 1973). Mountjoy and co-workers (1977) and Solyom *et al*. (1981) also found weak effects. Marks (1983) argued that benefit in phobic patients only occurred when depressive symptoms were present, but this was not the case in Tyrer's study.

Responsive subgroups

Atypical depression

With such a mix of positive and negative controlled trials it is not surprising that interest has focused on diagnostic subtypes which might preferentially be responsive. Sargant and his colleagues originally suggested that depressives characterized as 'atypical' responded. The first clear published description was by West and Dally (1959). In a retrospective analysis patients responding favourably to iproniazid were found to show absence of self-reproach, morning worsening and early wakening; and presence of evening worsening, hysterical symptoms, tremor, and a history of having been worsened by ECT. West and Dally went further and described a syndrome

Table 1 *Controlled trials of MAO inhibitors against placebo in depression*

Drug		Authors	Patient type	Source of sample
Phenelzine	Positive studies	Rees and Davies (1961)	Predominantly endogenous	In-patient
		Lascelles (1966)	Pain and depression	Out-patient
		Johnstone and Marsh (1973)	Neurotic	Out-patient
		Robinson et al. (1973)	Neurotic/atypical	Out-patient
		Ravaris et al. (1976)	Neurotic/atypical	Out-patient
		Raft et al. (1979)	Pain and depression	Mainly out-patient
		Paykel et al. (1982a)	Mixed: mainly neurotic	Out-patient
		Georgotas et al. (1987)	Elderly major depression	Out-patient
		Liebowitz et al. (1988)	Atypical	Out-patient and in-patient
		Quitkin et al. (1988)	Atypical	Out-patient and in-patient
	Doubtfully positive	Hutchinson and Smedberg (1960)	Atypical	In-patients
		Hare et al. (1962)	Mixed	Day patients
		Mountjoy et al. (1977)	Neurotic depressives, anxiety states, phobics	Mixed
	Negative studies	Greenblatt et al. (1964)	Mixed	In-patient
		Medical Research Council (1965)	Mixed	In-patient
		Bellak and Rosenberg (1966)	Mixed	In-patient
		Raskin et al. (1974)	Mixed	In-patient
Isocarboxazid	Positive studies	Joshi (1961)	Mixed	In-patient
		Kurland et al. (1967)	Mixed	In-patient
		Davidson et al. (1988)	Anxiety-Depression	Out-patient
	Negative studies	Rothman et al. (1962)	Mixed	In-patient
		Overall et al. (1962)	Mixed	In-patient
		Greenblatt et al. (1954)	Mixed	In-patient
Tranylcypromine	Positive studies	Bartholomew (1962)	Mixed	Out-patient
		White et al. (1984)	Predominantly neurotic	Out-patient
	Doubtfully positive	Khanna et al. (1983)	Mixed	In-patient
	Negative studies	Gottfries (1963)	Mixed	In-patient

shown by patients who had often been ill for years, had phobic anxiety, appeared anxious and overactive, were fatigued, had early morning wakening, and morning worsening. Some gave an impression of lifelong inadequacy or hysteria, although on close examination the premorbid personality was non-neurotic.

Dally and Rohde (1961) compared a small group of responders to imipramine with responders to iproniazid. The former contained a high proportion of endogenous depressives, the latter non-endogenous depressives. Later papers from St Thomas's Hospital laid more emphasis on anxiety states. Sargant and Dally (1962), in a clinical description, reported a good response in anxiety states to MAO inhibitors plus chlordiazepoxide. Kelly and co-workers (1970) reported retrospectively on a large series of phobic patients treated with MAO inhibitors with apparent good response, as shown by improvement in phobic ratings, depression, and panic attacks. Pollitt and Young (1971) found a good deal of overlap in the symptom patterns of anxious patients responding to MAO inhibitors and those of depressives.

When we came to look over the literature we were impressed that there were three different meanings which had been used in referring to atypical depression (Paykel *et al.* 1983). The first one was that of marked anxiety and phobic symptoms, either accompanied by depression, or assumed to have some relation to an underlying depression as indicated, for instance, by diurnal variation. This would include the influential views of Tyrer (1976).

The second meaning is what Pollitt (1965) describes as reversed functional shift, i.e. depression with a diurnal pattern of evening worsening, insomnia of early rather than late kind, or increased sleep, appetite, and weight, all in the direction opposite to the physiological changes said to characterize endogenous depression. In the USA, Klein and colleagues (Liebowitz *et al.* 1988) assign principal importance to these features.

A third meaning of atypical depression, at least in earlier studies, was non-endogenous depression in general. Robinson, Nies, and Ravaris employed a diagnostic index for typical and atypical depression (Robinson *et al.* 1974) which corresponds to this view.

These three meanings may identify different groups of patients. From a sample of out-patient depressives, we collected a large amount of data relevant to diagnosis including specific diagnoses on many different sub-classificatory systems (Paykel *et al.* 1983). We examined the relationships between different diagnostic systems using separate criteria. Groups corresponding to the separate concepts related poorly and tended to select quite different patients.

Predictors of response

There have been a few open studies employing statistical predictor methods. Some early studies found few predictive features (Paykel 1979).

Kay and colleagues (1973) reported a four-week comparative trial in depressed out-patients treated with amitriptyline or with rather low-dose phenelzine. There were few differences in overall outcome. Some specific features were found to predict outcome with one drug or the other in separate analyses of the two treatment groups, but these features were not very consistent with the usual views of responsive subtypes. In an open predictor study of four weeks with phenelzine we employed multiple predictors (Paykel *et al.* 1979). Better response was found in the following groups: out-patients and day patients, rather than in-patients; atypical depressives on Nies' diagnostic index; depressives characterized by milder illness with an admixture of anxiety; and in hostile depressives and agitated depressives in Overall's typology, with anxious depressives intermediate, and retarded depressives doing very much the worst.

Open predictor studies are not a good way of seeking responsive subgroups. The elements producing response in any group of patients given a single treatment are complex: they include spontaneous remission, non-specific effects of all other helping interventions which are occurring simultaneously, as well as effects of the specific treatment, and there is no way of telling which of these is contributing most to predicition. A better method is to undertake comparison with placebo and with another class of drug, to see which subgroups show most benefit in terms of superiority over alternative treatments and placebo (Paykel 1988).

The placebo-controlled trials enable some analysis of this, and details are also summarized in Table 1. In these studies there was not a very strong relationship between responses and defined subtype, but there tended to be a relationship to treatment setting. More of the studies showing drug–placebo differences have been in out-patient samples, and more of the negative studies have been carried out on in-patients. The most likely explanation would be that this reflects diagnostic subtype, since out-patients are likely to be less severely ill and more neurotic in symptoms pattern.

Comparison of MAOIs and tricyclics

Comparative trials of MAO inhibitors and tricyclic antidepressants offer another avenue of approach which is more complex, since the issue of differential responsiveness of tricyclics is also involved. Three of the large in-patient studies which found no difference between MAO inhibitors and placebo, found tricyclics superior to both (Greenblatt *et al.* 1964; MRC 1965; Overall *et al.* 1962). Some smaller early trials of MAO inhibitors against tricyclics failed to find clear differences overall or by subtype (Paykel 1979); so, more recently, did White *et al.* (1984). Vallejo and co-workers (1987) found phenelzine comparable to imipramine in major depression with melancholia, but superior in dysthymic disorder.

Some placebo-controlled trials have paid particular attention to classification. Robinson and co-workers (1974) used a diagnostic index, corresponding to a broad non-endogenous definition of atypical depression. Pooling patients from two studies they found a weak tendency for patients at the atypical pole to improve with phenelzine, but not with the placebo. In a direct comparison of phenelzine and amitriptyline by the same group of investigators (Ravaris *et al.* 1980) there were relatively few differences; phenelzine tended to have more effect on anxiety measures.

We carried out a controlled trial of phenelzine, amitriptyline, and placebo in out-patients with depression and mixed anxiety-depression (Rowan *et al.* 1982; Paykel *et al.* 1982*a*). Both drugs were superior to placebo and differences between them were small. Phenelzine tended to have better effect on anxiety measures and in the presence of initial anxiety: there were no differential effects in respect of either of the other two concepts of atypical depression. In this study amitriptyline was effective across a broad range of patients: phenelzine appeared more selective. Davidson and co-workers (1986), analysing patients from a number of studies, found only weak differences between patients responding to the two classes of antidepressant. There was a tendency for better effects of MAO inhibitor rather than tricyclic where a precipitant was present and, contrary to expectation, in the absence of agoraphobia. When separate analyses were carried out in the two sexes some further predictors were found, but these were not consistent between the sexes.

Liebowitz and colleagues (1988) found stronger differences. Selecting patients for atypical depression, defined by reversed functional shift and mood reactivity, they found 71 per cent of patients responding to phenelzine, 50 per cent to imipramine, and 28 per cent to placebo. Examining subgroups within this selected sample, they unexpectedly found that superiority of both phenelzine and imipramine to placebo was largely confined to patients when spontaneous panic attacks and/or hysteroid dysphoric features. Davidson and colleagues (1988), in a comparative trial of isocarboxazid and placebo, found the drug superior to placebo in major but not minor depressives and significantly more effective in depression classified as endogenous or its DSM III equivalent, melancholia, by various diagnostic criteria. Somewhat inconsistent with this, the drug was more effective than placebo in atypical depression with reversed functional shift and in Overall's subtypes of anxious and hostile depression. In general, these two recently published studies do suggest a little more selectivity for reversed functional shift than the earlier studies, and some convergence with presence of anxiety, but still the overall impression is of weak differences. Even the conclusion that in-patients do not benefit may need re-evaluation, since most of the in-patient studies were early ones, using doses and treatment periods which were relatively low by modern standards. Many clinicians use MAO inhibitors with benefit on in-patients who have failed to respond to other treatments, and this may extend to bipolar depressives

The efficacy of tricyclic antidepressants

Proper conclusions about MAOI-responsive subgroups also depend on the demonstration that tricyclic antidepressants are not effective in these patients, and here there is room for considerable doubt.

The view that tricyclics are particularly effective in endogenous depression has an impressive lineage, going back as early as the initial studies. However, a critical look at the controlled trials published in the 1960s suggests that, although the effect may have been a little better in endogenous depressives, there was also evidence of benefit in neurotic depressives. Certainly there have been many controlled trials in which a tricyclic was superior to placebo in samples characterized as neurotic or reactive (Paykel 1979). We recently found amitriptyline superior to placebo in general practice depressives irrespective of endogenous symptoms (Paykel *et al*. 1988). Only a relatively low threshold of severity determined therapeutic benefit.

More recent findings regarding severe and delusional depression also present a problem. There have been a number of reports indicating that delusions predict a poor response to tricyclic antidepressants and a better response to ECT (Paykel 1979). It would appear that the most severe endogenous or psychotic depressives do not respond very well to tricyclics.

There is increasing evidence that tricyclics are superior to placebo in neurotic states other than depression. Marks (1983) in his review drew attention to several studies showing tricyclics superior to placebo in obsessional neurosis and in agoraphobia. The obsessional studies depend mainly on one drug, clomipramine, with a more serotoninergic action, but the studies in agoraphobia and panic disorder have employed more standard drugs. Sheehan and co-workers (1980) also included a tricyclic antidepressant and found it better than placebo, although a little inferior to the MAO inhibitor.

A reasonable conclusion is that there are probably some differences from MAO inhibitors but they are weak. Tricyclics appear relatively broad-spectrum antidepressants, with the best effect in the presence of endogenous symptoms (Rao and Coppen 1978) without great severity or delusions; MAO inhibitors appear preferentially effective in the presence of anxiety and reversed functional shift, and may be less broadly effective, but the overlap is great.

Predictors other than clinical subtype

History of response

It is possible that there is some variable which does select MAO inhibitor responders, but that it is biochemical and only weakly reflected in clinical classification. One piece of evidence comes from previous history of response. A retrospective comparison of treatments (Pare and Mack 1971) suggested that when depressives were given a second tricyclic or MAO inhibitor in a subsequent episode, response to drugs of the same type was similar, but across classes it was inconsistent. There is, however, one negative study (El-Islam 1973). Hamilton (1982) administered ECT to patients who had randomly received phenelzine or imipramine and failed to respond. Patients who had not responded to imipramine had significantly worse response to ECT than those who had unsuccessfully received phenelzine, suggesting that imipramine responders and ECT responders are similar, but different to phenelzine responders. Clinically it is not uncommon to find patients showing consistency of response in episodes and this is often used as a guide to treatment choice.

Response within families

Genetic responsiveness has been reported. Pare and colleagues (1962) reported 12 patient–relative correlations, some involving more than one drug, in which patient and relative received the same type of drug, and in each the response was similar. In ten pairs, unlike drugs were received: in six of these there was also concordance, with no improvement in either class; in four, improvement occurred on one class but not on the other; in no case was there improvement in a pair on drugs of different classes (Pare and Mack 1971). In a further sample of 11 like-treatment pairs, nine were concordant, and eight unlike-treatment pairs, seven were discordant (Pare and Mack 1971). Angst (1961) presented results in 47 patient–relative pairs with endogenous depression treated with imipramine, with high concordance. However, the majority showed a good response; the possibility of non-drug related improvement and the absence of a comparison with MAO inhibitors render these findings less convincing.

Biochemical predictors

A biochemical predictor has also been isolated, depending on polymorphism for the enzyme liver-acetyl-transferase. About 60 per cent of the British population are slow acetylators and 40 per cent fast acetylators. Hydrazines are substrates of this enzyme: slower inactivation of hydrazine MAO inhibitors could lead to greater efficacy. Published studies have involved phenelzine. Evans and co-workers (1965) failed to show evidence of greater efficacy in slow acetylators in an open study, but did find more side-effects. Johnstone and Marsh (1973) found drug–placebo differences in slow but not in fast acetylators in a controlled trial of phenelzine 90 mg daily for three weeks. Johnstone (1976), in an open study, found greater inhibition of MAO as reflected in urinary-tryptamine excretion, and also greater urinary free-phenelzine excretion, suggesting less acetylation. Paykel and co-workers (1982*b*) found more improvement and greater phenelzine–

placebo differences in slow acetylators. However no effects were found in other studies variously examining improvement, side effects, and degree of MAO inhibition (Davidson *et al.* (1978); Marshall *et al.* (1978); Yates and Loudon (1979); and Tyrer *et al.* (1980)).

Overall, there appears to be an effect, but not a very strong one. It is not clear whether this is the gene responsible for consistency in family studies. In any event, the effect is unlikely to be a fundamental one related to depressive disorder: it involves a pharmacokinetic mechanism producing slow or rapid inactivation. Clinically, rather than testing acetylator phenotype it is easier and probably just as effective to increase the dose until dose-related side-effects indicate adequate levels.

Effects on serotonin

A further, more speculative, biochemical indicator could involve serotoninergic mechanisms. Clomipramine is particularly effective in obsessional disorders and is more serotoninergic than other, older tricyclics. There is some evidence that the newer serotonin re-uptake inhibitors may be more effective in neurotic and anxious pictures (Norton *et al.* 1984; Evans *et al.* 1988; Den Boer and Westenberg 1988). MAO inhibitors may be serotinergic: at least their interaction with tricyclics resembles the serotonin syndrome, and clomipramine is particularly liable to produce it. This speculation would be consistent with the anxiolytic effects of $5HT1_A$ agonists such as buspirone.

Clinical guidelines

In spite of all the research, guidelines are not yet clear. The most useful that can be suggested appear fairly blunt.

1. *First choice*   Given the greater ease of use of alternative treatments there only appears to be one situation in which clinicians would regard MAO inhibitors as a first choice: where there is a clear prior history of previous good response, and the clinical picture of that episode was not grossly different to the present one. In most other circumstances in treatment of depression and anxiety, if an antidepressant is to be used, the greater safety and comparable efficacy of tricyclics make them a better choice. The exception is in very severe depression, particularly with delusions and retardation, where ECT is often a first choice.

2. *Second choice*   If the first choice fails, guidelines for second choice have not usually been formulated but severity and clinical picture make reasonable criteria. Where a depressive disorder is severe, particularly if the symptom picture is endogenous, ECT would appear a better second choice; where it is less severe, particularly where there is anxiety or reversed functional shift, an MAO inhibitor would appear to be

indicated. In anxiety disorders ECT is not indicated: MAO inhibitors are a strongly ranking second choice where an antidepressant is used.

3. *Third choice*   Where the second choice has not been effective in the treatment of depression, MAO inhibitors, if not used already, are a reasonable third choice.

4. *Combinations*   The above choices all refer to single therapy. There is now good evidence that lithium potentiates tricyclics (Paykel and Van Woerkom 1987) and probable evidence that it potentiates MAO inhibitors (Fein *et al.* 1988). Many would regard addition of lithium as an easy second step after failure of first drug choice, before changing anti-depressants, although this is not yet a fully established pattern.

   Other combinations have been used. Two major trials have failed to show MAOI-tricyclic combinations superior to the best of the two single constituents (Young *et al.* 1979; Razani *et al.* 1983) but neither was in resistant cases and anecdotal evidence suggests that the combination is occasionally of benefit. Four studies have found tryptophan to potentiate MAO inhibitors, with one study failing to do so (see Paykel and Van Woerkom 1987). Such specialized treatment choices which have an occasional place in resistant depression, are reviewed by Goodwin (Chapter 1).

5. *Choice of MAO inhibitor*   There have been few comparative trials of different MAO inhibitors. Clinical experience suggests that tranyl-cypromine is more potent and has more amphetamine-like effect, but that at high dose there is little to distinguish the different MAO inhibitors. Iproniazid is potentially hepatotoxic and I believe that its use is no longer appropriate. The probable increase in cheese reactions with tranylcypro-mine, and its potency, would suggest that it might be more appropriate for in-patient, resistant depressives, and phenelzine or isocarboxazid more for out-patients. The arguments are not very strong in either direction, but this is a common clinical practice, and is my own also.

References

Angst, J. (1961). A clinical analysis of the effects of tofranil in depression. Longi-tudinal and follow-up studies. Treatment of blood-relations. *Psychopharmaco-logia*, **2**, 381–407.

Bartholomew, A. A. (1962). An evaluation of tranylcypromine (Parnate) in the treatment of depression. *Medical Journal of Australia*, **49**, 655–62.

Bellak, L. and Rosenberg, S. (1966). Effects of anti-depressant drugs on psycho-dynamics. *Psychomatics*, **vii**, 106–14.

Blackwell, B. (1963). Tranylcypromine (letter). *Lancet*, **i**, 414.

Dally, P. J. and Rohde, P. (1961). Comparison of antidepressant drugs in depress-ive illness. *Lancet*, **i**, 18–20.

Davidson, J., McLeod, M. N., and Blum, M. (1978). Acetylation phenotype, platelet monoamine oxidase inhibition and the effectiveness of phenelzine in depression. *American Journal of Psychiatry*, **135**, 467–9.

Davidson, J., Pelton, S., Krishnan, R. R., and Allf, B. (1986). The Newcastle Anxiety Depression Index in relationship to the effects of MAOI and TCA drugs. *Journal of Affective Disorders*, **11**, 51–61.

Davidson, J. R. T., Giller, E. L., Zisook, S., and Overall, J. E. (1988). An efficacy study of isocarboxazid and placebo in depression, and its relationship to depressive nosology. *Archives of General Psychiatry*, **45**, 120–7.

Den Boer, J. A. and Westenberg, H. G. M. (1988). Effect of a serotonin and noradrenaline uptake inhibitor in panic disorders. *International Clinical Psychopharmacology*, **3**, 59–74.

El-Islam, M. F. (1973). Is response to antidepressants an aid to differentiation of response-specific types? *British Journal of Psychiatry*, **123**, 509–11.

Evans, D. A., Davison, K., and Pratt, R. T. C. (1965). The influence of acetylator phenotype on the effects of treating depression with phenelzine. *Clinical Pharmacology and Therapeutics*, **6**, 430–5.

Evans, L., Kenardy, J., Schneider, P., and Hoey, H. (1986). Effect of a selective serotonin uptake inhibitor in agoraphobia with panic attacks. A double-blind comparison of zimeldine, imipramine and placebo. *Acta Psychiatrica Scandinavica*, **73**, 49–53.

Fein, S., Paz, V., Rao, N., and LaGrassa, J. (1988). The combination of lithium carbonate and an MAOI in refractory depressions. *American Journal of Psychiatry*, **145**, 249–50.

Georgotas, A., McCue, R. E., Friedman, E., and Cooper, T. B. (1987). Response of depressive symptoms to nortriptyline, phenelzine and placebo. *British Journal of Psychiatry*, **151**, 102–6.

Gottfries, C. G. (1963). Clinical trial with the monoamino oxidase inhibitor tranylcypromine on a psychiatric clientele. *Acta Psychiatrica Scandinavica*, **39**, 463–72.

Greenblatt, M., Grosser, G. H., and Wechsler, H. (1964). Differential response of hospitalised depressed patients to somatic therapy. *American Journal of Psychiatry*, **120**, 935–43.

Hamilton, M. (1982). The effect of treatment on the melancholias (depressions). *British Journal of Psychiatry*, **140**, 223–30.

Hare, E. H., Dominian, J., and Sharpe, L. (1962). Phenelzine and dexamphetamine in depressive illness. A comparative trial. *British Medical Journal*, **1**, 9–12.

Hutchinson, J. T. and Smedberg, D. (1960). Phenelzine ('Nardil') in the treatment of endogenous depression. *Journal of Mental Science*, **106**, 704–10.

Johnstone, E. C. (1976). The relationship between acetylator status and inhibition of monoamine oxidase, excretion of free drug and antidepressant response in depressed patients on phenelzine. *Psychopharmacologia (Berl.)*, **46**, 289–94.

Johnstone, E. C. and Marsh, W. (1973). Acetylator status and response to phenelzine in depressed patients. *Lancet*, **i**, 567–70.

Joshi, V. G. (1961). Controlled clinical trial of isocarboxazid (Marplan) in hospital psychiatry. *Journal of Mental Science*, **107**, 567–71.

Kay, D. W. K., Garside, R. F., and Fahy, T. J. (1973). A double-blind trial of phenelzine and amitriptyline in depressed outpatients. A possible differential effect of the drugs on symptoms. *British Journal of Psychiatry*, **123**, 63–7.

Kelly, D., Guirguis, W., Frommer, E., Mitchell-Heggs, N., and Sargant, W. (1970). Treatment of phobic states with antidepressants: a retrospective study of 246 patients. *British Journal of Psychiatry*, **116**, 387–98.

Khanna, J. L., Pratt, S., Burdizk, E. G., and Chaddha, R. L. (1983). A study of certain effects of tranylcypromine, a new antidepressant. *Journal of New Drugs*, **3**, 227–32.

Kurland, A. A., Destounis, N., Shaffer, J. W., and Pinto, A. (1967). A critical study of isocarboxazid (Marplan) in the treatment of depressed patients. *Journal of Nervous and Mental Disease*, **145**, 292–305.

Lascelles, R. G. (1966). Atypical facial pain and depression. *British Journal of Psychiatry*, **112**, 651–59.

Liebowitz, M. R., *et al.* (1988). Antidepressant specificity in atypical depression. *Archives of General Psychiatry*, **45**, 129–37.

Lipsedge, B. S., Hajioff, J., Huggins, P., Napier, L., Pearce, J., Pike, D. J., and Rich, M. (1973). The management of severe agoraphobia: a comparison of iproniazid and systemic desensitization. *Psychopharmacologia*, **32**, 67–80.

Marks, I. (1983). Are there anticompulsive or antiphobic drugs? Review of the evidence. *British Journal of Psychiatry*, **143**, 338–47.

Marshall, E. F., Mountjoy, C. Q., Campbell, I. C., Garside, R. F., Leitch, I. M., and Roth, M. (1978). The influence of acetylator phenotype on the outcome of treatment with phenelzine in a clinical trial. *British Journal of Clinical Pharmacology*, **6**, 247–54.

Medical Research Council (1965). Clinical trial of the treatment of depressive illness. *British Medical Journal*, **1**, 881–6.

Morris, J. B. and Beck, A. T. (1974). The efficacy of antidepressant drugs: a review of research (1958–1972). *Archives of General Psychiatry*, **30**, 667–74.

Mountjoy, C., Roth, M., Garside, R. F., and Leitch, I. A. (1977). A clinical trial of phenelzine in anxiety and depressive and phobic neurosis. *British Journal of Psychiatry*, **131**, 486–92.

Norton, K. R. W., Sireling, L. I., Bhat, A. V., Rao, B., and Paykel, E. S. (1984). A double-blind comparison of fluvoxamine, imipramine and placebo in depressed patients. *Journal of Affective Disorders*, **7**, 297–308.

Overall, J. E., Hollister, L. E., Pokorny, A. D., and Casey, J. F. (1962). Drug therapy in depressions. Controlled evaluation of imipramine, isocarboxazide, dextroamphetamine-amobarbital, and placebo. *Clinical Pharmacology and Therapeutics*, **3**, 16–22.

Pare, C. M. B. and Mack, J. W. (1971). Differentiation of two genetically specific types of depression by the response to antidepressant drugs. *Journal of Medical Genetics*, **8**, 306–9.

Pare, C. M. B., Rees, L., and Sainsbury, M. J. (1962). Differentiation of two genetically specific types of depression by the response to antidepressants. *Lancet*, **2**, 1340–3.

Paykel, E. S. (1979). Predictors of treatment response. In *Psychopharmacology of affective disorders* (ed. E. S. Paykel and A. Coppen), pp. 193–220. Oxford University Press.

Paykel, E. S. (1988). Antidepressants: their efficacy and place in therapy. *Journal of Psychopharmacology*, **2**, 105–8.

Paykel, E. S. and Van Woerkom, A. E. (1987). Pharmacologic treatment of resistant depression. *Psychiatric Annals*, **17**, 327–31.

Paykel, E. S., Parker, R. R., Penrose, R. J., and Rassaby, E. (1979). Depressive classification and prediction of response to phenelzine. *British Journal of Psychiatry*, **134**, 572–81.

Paykel, E. S., Rowan, P. R., Parker, R. R., and Bhat, A. V. (1982*a*). Response to phenelzine and amitriptyline in subtypes of outpatient depression. *Archives of General Psychiatry*, **39**, 1041–9.

Paykel, E. S., West, P. S., Rowan, P. R., and Parker, R. R. (1982*b*). Influence of acetylator phenotype on antidepressant effects of phenelzine. *British Journal of Psychiatry*, **141**, 243–8.

Paykel, E. S., Parker, R. R., Rowan, R. R., Rao, B. M., and Taylor, C. N. (1983). Nosology of atypical depression. *Psychological Medicine*, **13**, 131–9.

Paykel, E. S., Hollyman, J. A., Freeling, P., and Sedgwick, P. (1988). Predictors of therapeutic benefit from amitriptyline in mild depression: a general practice placebo-controlled trial. *Journal of Affective Disorders*, **14**, 83–95.

Pollitt, J. (1965). *Depression and its treatment*. Heinemann, London.

Pollitt, J. and Young, J. (1971). Anxiety state or masked depression? A study based on the action of monoamine oxidase inhibitors. *British Journal of Psychiatry*, **119**, 143–9.

Quitkin, F. M., *et al.* (1988). Phenelzine versus imipramine in the treatment of probably atypical depression: defining syndrome boundaries of selective MAOI responders. *American Journal of Psychiatry*, **145**, 306–11.

Raft, D., Davidson, J., Mattox, A., Mueller, R., and Wakik, J. (1979). Double-blind evaluation of phenelzine, amitriptyline, and placebo in depression associated with pain. In *Monoamine oxidase: structure, function, and altered functions* (ed. T. P. Singer, R. W. Von Korff, and D. L. Murphy), pp. 507–16. Academic Press, New York.

Rao, V. A. and Coppen, A. (1978). Classification of depression and response to amitriptyline therapy. *Psychological Medicine*, **9**, 321–5.

Raskin, A., Schulterbrandt, J. G., Reatig, N., Crook, T. H., and Odle, D. (1974). Depression subtypes and response to phenelzine, diazepam and a placebo. *Archives of General Psychiatry*, **30**, 66–75.

Ravaris, C. L., Nies, A., Robinson, D. S., Ives, J. O., Lamborn, K. R., and Korson, L. (1976). A multiple-dose, controlled study of phenelzine in depression-anxiety states. *Archives of General Psychiatry*, **33**, 34–50.

Ravaris, C. L., Robinson, D. S., Ives, J. O., Nies, A., and Bartlett, D. (1980). Phenelzine and amitriptyline in the treatment of depression. *Archives of General Psychiatry*, **37**, 1076–80.

Razani, J., *et al.* (1983). The safety and efficacy of combined amitriptyline and tranylcypramine antidepressant treatment. *Archives of General Psychiatry*, **40**, 657–61.

Rees, L. and Davies, B. (1961). Controlled trial of phenelzine (Nardil) in treatment of severe depressive illness. *Journal of Mental Science*, **107**, 560–6.

Robinson, D. S., Nies, A., Ravaris, C. L., Ives, J. O., and Lamborn, K. R. (1974). Treatment response to MAO inhibitors: Relation to depressive typology and blood platelet MAO inhibition. In *Classification and prediction of outcome of depression*, (ed. J. Angst), pp. 259–67. F.K. Shattauer Verlag, Stuttgart.

Robinson, D. S., Nies, A., Ravaris, L. C. and Lambourn, K. R. (1973). The monoamine oxidase inhibitor, phenelzine, in the treatment of depressive-anxiety states: a controlled clinical trial. *Archives of General Psychiatry*, **29**, 407–13.

Rothman, T., Grayson, H., and Ferguson, J. (1962). A comparative investigation of isocarboxazid and imipramine in depressive syndromes. A second report. *Journal of Neuropsychiatry*, **3**, 324–40.

Rowan, P. R., Paykel, E. S., and Parker, R. R. (1982). Phenelzine and amitriptyline: effects on symptoms of neurotic depression. *British Journal of Psychiatry*, **140**, 475–83.

Sargant, W. and Dally, P. (1962). Treatment of anxiety states by antidepressant drugs. *British Medical Journal*, **1**, 5–9.

Sheehan, D., Ballenger, J., and Jacobsen, G. (1980). Treatment of endogenous anxiety with phobic, hysterical and hypochondriacal symptoms. *Archives of General Psychiatry*, **36**, 51–9.

Solyom, L., Heseltine, G. F. D., McClure, D. J., Solyom, C., Ledwidge, B., and Steinberg, G. (1973). Behaviour therapy versus drug therapy in the treatment of phobic neurosis. *Canadian Psychiatric Association Journal*, **18**, 25–32.

Solyom, C., Solyom, L., LaPierre, Y., Pecknold, J., and Morton, L. (1981). Phenelzine and exposure in the treatment of phobias. *Biological Psychiatry*, **16**, 239–247.

Tyrer, P. (1976). Towards rational therapy with monoamine oxidase inhibitors. *British Journal of Psychiatry*, **128**, 354–360.

Tyrer, P., Candy, J., and Kelly, D. (1973). A study of the clinical effects of phenelzine and placebo in the treatment of phobic anxiety. *Psychopharmacologia*, **32**, 237–54.

Tyrer, P., Gardner, M., Lambourn, J., and Whitford, M. (1980). Clinical and pharmacokinetic factors affecting response to phenelzine. *British Journal of Psychiatry*, **136**, 359–65.

Vallejo, J., Gasto, C., Catalan, R., and Salamero, M. (1987). Double-blind study of imipramine versus phenelzine in melancholias and dysthymic disorders. *British Journal of Psychiatry*, **151**, 639–42.

West, E. D. and Dally, P. J. (1959). Effects of iproniazid in depressive syndromes. *British Medical Journal*, **1**, 1491–9.

White, K., Razani, J., Cadow, B., Gelfand, R., Palmer, R., Simpson, G., and Sloane, R. B. (1984). Tranylcypromine vs. nortriptyline vs. placebo in depressed outpatients: a controlled trial. *Psychopharmacology*, **82**, 258–62.

Yates, C. M. and Loudon, J. B. (1979). Acetylator status and inhibition of platelet monoamine oxidase following treatment with phenelzine. *Psychological Medicine*, **9**, 777–9.

Young, J. P. R., Lader, M. H., and Hughes, W. C. (1979). Controlled trial of trimipramine, monoamine oxidase inhibitors and combined treatment in depressed outpatients. *British Medical Journal*, **279**, 1315–17.

3

Mania: what if neuroleptics don't work?

JULIE CHALMERS

The problem

Neuroleptic drugs are the mainstay of the treatment of patients admitted to hospital suffering from acute mania. In general, treatment with neuroleptics is effective in approximately 70 per cent of patients (Peet 1975).

Two main problems arise if these drugs do not have the desired effect. The most common problem is one of partial response, where motor activity is reduced and sleep is improved but manic ideation continues. The problem of total non-response is rare, but leads to major clinical problems. This chapter will focus mainly on treatment approaches to the latter, more difficult problem.

Unfortunately, because information from the literature is extremely sparse, no clear guidelines to the next step in management exist. Controlled pharmacological studies present major problems in severely manic patients due to difficulties in gaining adequate co-operation, and the possible need for additional medication. Information is therefore based on anecdotal reports or from studies of less disturbed patients.

Before concluding that a patient is neuroleptic-resistant it is worth considering two areas, compliance and diagnosis. The question of adequate compliance, particularly if oral neuroleptics are used, should be fully explored. Liquid preparations or intramuscular injections may help overcome this problem.

Attention should also be given to the possibility that the mania is secondary to an organic condition, particularly if the patient fails to respond to conventional treatment, or appears to be responding in an unusual way, for example, with undue sensitivity, or has any symptoms classically associated with organic states (see Krauthammer and Klerman 1978). The possibility of HIV infection should also be remembered (see Chapter 17).

The environment in which a patient is nursed may have an effect on his or her level of behavioural disturbance (Bouras et al. 1982). A balance between over-stimulation and a worsening of irritability and frustration due to confinement in a single room must be achieved if drugs are to have their optimal effect.

Staff attitudes are also important. It is vital that the sense of hopelessness experienced by the staff in the face of non-response is not transmitted to the patient or the patient blamed for not getting better.

Medication

Neuroleptic dose

Before concluding that neuroleptic treatment has been ineffective it is necessary to consider whether it has been given in sufficient dose. What constitutes adequate dose is a matter of some debate. Studies of groups of patients suggest that extremely high doses are not only unwarranted but may even lead to a worsening of the clinical picture (Baldesarrini *et al.* 1988). However, clinical experience would suggest that there are occasionally patients who may benefit from high-dose neuroleptic treatment.

Changing neuroleptics

In the face of apparent neuroleptic non-response some clinicians may consider a change in class of neuroleptic drug or a change in the form of administration, such as the use of depot-neuroleptic medication. There is no evidence to suggest that any particular neuroleptic drug has a specific impact on manic symptomatology and no evidence to support the view that depot preparations are superior to other methods of administration. A study by Shopsin and coworkers (1975) comparing haloperidol, chlorpromazine, and lithium suggested that haloperidol had a more rapid onset of action and a greater effect on motor activity than the other two drugs. These actions are particularly welcome in the management of mania and may partly explain haloperidol's popularity in the treatment of this disorder. Chlorpromazine, with its α-adrenergic blocking properties, is more likely to cause sedation and postural hypotension, perhaps limiting its use in sufficient dose. Droperidol, a butyrophenone derivative, is a potent neuroleptic drug with greater sedative activity than haloperidol. It has been suggested to be of particular benefit in excited states but there is no convincing evidence at present to support its use over other drugs in the treatment of mania.

Lithium

Lithium has been clearly demonstrated to be an antimanic agent in the acute phase of disturbance in placebo-controlled trials and in comparative trials against antipsychotic medication (Goodwin and Zis 1979). Compared to lithium, neuroleptics have a more rapid onset of action and give rise to more sedation and decrease in motor activity. Lithium takes longer to exert its effect, up to one to two weeks, but has a greater impact on the core disturbance in terms of normalizing mood, manic behaviour, and ideation.

Carbamazepine

Carbamazepine is an anticonvulsant drug with particular efficacy in the treatment of temporal lobe epilepsy. Post and co-workers (1982) have suggested that carbamazepine's effect on limbic structures may account for its effectiveness in the treatment of affective disturbance unrelated to seizure activity.

Both open and double-blind studies suggest that carbamazepine may have an antimanic effect (Post 1982). Two groups of patients have been studied; those who have received no prior treatment during the index episode and a second group comprising patients who have failed to respond to conventional treatment (usually lithium non-responders).

There are some clinical pointers to patients who may specifically benefit from carbamazepine. Those with a history of a poor response to lithium and those with several episodes of mood disturbance, so-called rapid cycling (Dunner and Fieve 1974), may experience particular benefit.

Several practical aspects of the use of carbamazepine need emphasizing. Most patients are managed on between 800–1200 mg carbamazepine per day. Since patients may experience some nausea on first taking this drug the dose should be gradually increased over four to six days. Dizziness, drowsiness, and ataxia occur in 15–25 per cent of patients. Rashes also occur in approximately 10 per cent and may be harbingers of more serious toxic effects. In the initial stages of treatment, full blood count (carbamazepine may cause blood dyscrasias), urea and electrolytes (it may also cause hyponatraemia), and liver function (cholestatic jaundice has been reported) should be checked at regular intervals. As carbamazepine induces liver metabolism, patients on the low-dose oral contraceptive should be given a higher oestrogen-containing compound or be changed to non-hormonal methods. This interaction may be particularly pertinent in the overactive, sexually disinhibited female manic patient.

Clonazepam

In a double-blind crossover study of twelve acutely ill manic patients (Chouinard *et al.* 1983), clonazepam in doses ranging from 4–16 mg was found to be superior to lithium in the control of two symptoms, motor activity and over-talkativeness. However, it remains unclear whether the apparent improvement is due to the sedative effect alone rather than any specific antimanic action. It should be noted that the level of manic disturbance in this study was not severe.

As clonazepam is a benzodiazepine there is a small risk of paradoxical disinhibition. Clinical experience suggests this occurs rarely but may account for the occasional apparent worsening of symptoms when clonazepam is introduced.

Verapamil

Verapamil, a drug used in the management of angina and arrhythmias, has also been successfully used in the treatment of acute mania (Dubovsky *et al.* 1986). In a double-blind comparison (Giannini *et al.* 1986) with lithium there were no differences between drugs and both were superior to placebo in the treatment of mild to moderate mania. There was also a suggestion that verapamil had a greater impact than lithium on the symptoms of anxiety, tension, and excitement.

Since there are reports that verapamil may increase lithium excretion, serum levels should be monitored closely if the two drugs are used together. Combinations of carbamazepine and verapamil have led to cases of neurotoxicity (Jefferson *et al.* 1987, p. 164).

Clonidine

Clonidine, a well-established antihypertensive drug, has been used in the treatment of mania. Case reports from both the USA and Great Britain (Maguire and Singh 1987) suggest that clonidine, in doses ranging from 0.2–0.4 mg, may in combination with other drugs be helpful in reducing manic symptoms in treatment-resistant patients. The unsatisfactory nature of such case reports does not clearly indicate whether clonidine really does have an antimanic effect and this drug will require fuller evaluation. Initial pilot studies of clonidine versus other antimanic agents (Giannini *et al.* 1986) are inconclusive.

Sodium valproate

This anticonvulsant drug has been used in a small number of patients with some success. Fuller evaluation of its role in treatment and selection of appropriate patients is awaited.

Sedatives

Sometimes the need for behavioural control becomes urgent in the management of the severely disturbed manic patient. Although the following drugs play little part (with the possible exception of clonazepam) in treating the underlying disorder, they may have a role to play in bringing about rapid sedation.

Short-acting barbiturates such as amylobarbitone sodium may be given either orally or in the case of extreme urgency, intravenously. Caution is required in both acute administration and in withdrawal (see British National Formulary 1981). In some centres barbiturates are being used in preference to increasing doses of neuroleptic where sedation is thought to be desirable (G. Goodwin, personal communication).

Despite the statement in the British National Formulary that the use of paraldehyde as a sedative is obsolete, clinical experience suggests that doses of 10–20 ml given intramuscularly may be of considerable value in sedating a severely disturbed patient.

Clonazepam in large doses (up to 16 mg) may act as a potent sedative.

Combination treatment

Lithium plus neuroleptics: potential benefits or hazardous combination?

Clinically the combination of lithium and neuroleptics has been widely used. However, in the early 1970s reports suggested that this combination may give rise to an increase in lithium side-effects (Baldessarini and Stephens 1970). It should be noted, however, that in the 1970s lithium was likely to be maintained at much higher blood levels than are used today. The case reports of Cohen and Cohen (1974) describing irreversible brain damage in patients taking lithium and haloperidol together raised the spectre of a potentially dangerous drug interaction. Since then, further case reports have been published, many implicating the combination of haloperidol and lithium as a particularly hazardous one. However, the possibility that toxic levels of lithium alone were responsible in many cases has not been excluded. In addition, other drugs, such as anticholinergics which are commonly prescribed when potent neuroleptics are used, have also been suggested to play a part in the development of neurotoxicity.

Two groups of problems appear to exist. In some cases there is an exacerbation of the toxic effects of lithium. In others a picture similar to neuroleptic malignant syndrome, with rigidity, extrapyramidal symptoms, and autonomic disturbance, appears to occur. As with most adverse effects, the elderly may be at increased risk.

The consensus of opinion would now suggest that, apart from idiosyncratic reactions, the combination of neuroleptics and lithium is generally a safe one as long as lithium levels are kept within the current therapeutic range (not exceeding 0.8 mmol/L at 12 hours post dose). However caution must be exercised when high doses of potent neuroleptics are used.

Clinically the timing of the addition of lithium to neuroleptics can be problematic. Before introducing lithium the neuroleptic dose should be gradually reduced to the minimum amount which maintains the clinical improvement. The cautious 'safe' dose of haloperidol (20 mg/day) espoused by Schou (1986) is often, in clinical practice, found to be too low. A reduction to 40 mg of haloperidol before the introduction of lithium may be more clinically applicable. These figures are arbitrary and should be adjusted down if the patient is elderly or frail. The patient should be monitored closely for any increasing extrapyramidal symptoms. The use of anticholinergics should also be reduced to a minimum several days before the introduction of lithium.

With these precautions in mind is there any benefit in combining treatment? Unfortunately the literature has very little to offer on this topic.

Shopsin and coworkers (1975) briefly reported that in patients who failed initially to respond fully to neuroleptics, the addition of lithium led to a normalization in mood. This supports the common-sense view that if the drugs have slightly different actions then there may be benefit in using both together in the patient who fails to respond to single therapy.

Neuroleptics plus carbamazepine

A controlled double-blind study suggested that the combination of carbamazepine and haloperidol is superior to treatment with haloperidol alone in excited manic states (Klein *et al.* 1984). This study included patients who had a history of non-response to lithium and neuroleptics, further supporting the notion that carbamazepine may be of value in more atypical drug-resistant patients.

Carbamazepine plus lithium

There is anecdotal evidence from case reports (Lipinski and Pope 1982; Folks *et al.* 1982; Moss and James 1983) that the combination of lithium and carbamazepine may be helpful in patients who have failed to respond to neuroleptics or to lithium and carbamazepine used singly. There is also a hint that this combination is clinically more effective than neuroleptics and lithium together in this group of patients. However, considerable reservations have to be expressed as these case reports represent a highly selected and only moderately disturbed, atypical group of patients. In addition there have been reports of a possible interaction between carbamazepine and lithium giving rise to neurotoxicity (Chaudhry and Waters 1983).

Potential clinical problems with drug treatments

Compliance

All drugs (with the exception of some of the sedatives) are only available or used in oral form for the treatment of acute mania. Maggs (1963) demonstrated that a third of manic patients admitted to hospital are too disturbed to accept oral medication. This figure is likely to be even greater in patients who are severely disturbed.

Time lag of effect

Even if patients are willing to accept oral medication there may be a time lag of up to one week before drugs such as lithium take effect. In the management of a severely disturbed patient such a time lag may be unacceptable.

Possible interactions

All drugs mentioned have potentially serious side-effects which are often worsened when they are used in combination.

Electroconvulsive therapy (ECT)

ECT: evidence for efficacy in the management of mania

ECT was widely used in the treatment of mania in the 1940s prior to the introduction of lithium and neuroleptics in the 1950s. Fink (1977), in an extensive review of the early literature, suggested that ECT was thought to be an effective and safe treatment although this information came from case reports and open studies. The method of producing a seizure was different to that used in current clinical practice in that drugs such as pentylene tetrazole were often used, sometimes with only limited success in seizure induction.

Retrospective evaluation (McCabe 1976; McCabe and Norris 1977; Thomas and Reddy 1982) comparing matched groups of patients from different treatment eras have suggested that ECT is as effective as lithium and chlorpromazine. However, certain measures of outcome were crude, e.g., length of admission, and could be influenced by non-drug-related variables. A recent retrospective study by Black and coworkers (1987) implied that ECT may be particularly helpful in patients who have failed to respond to lithium and neuroleptics. Retrospective studies are, of course, methodologically flawed but all four studies show trends to suggest that ECT may be of some benefit in the treatment of acute mania.

There has been one prospective study (Small *et al.* 1988) of lithium versus ECT in the management of acute manic states. ECT was given three times weekly and lithium levels maintained between 0.6 and 1.5 mmol/L. For ethical and practical reasons neuroleptics were available for use if clinically indicated but the amount prescribed was kept to a minimum. Patients were not blind to treatment as sham ECT or placebo-lithium tablets were not used. Because of possible clues to treatment status the blindness of the raters was also compromised. To compensate for this, videotaped interviews in which references to treatment were omitted were also rated independently. Obviously conclusions from this study can only be tentative. However, ECT was shown to be an effective treatment and on many ratings, such as irritability, manic ideation, impulsivity, it was found to be superior to lithium over the first eight weeks. This difference was not accounted for by differences in the need for neuroleptic medication as there were no significant differences between groups.

Clinical experience of using ECT: some issues

Consent   The administration of ECT to a severely manic patient who is unable to give informed consent is identical to the situation faced by the clinician dealing with a severely depressed patient. Steps should therefore be taken to give the treatment under the Mental Health Act with a second opinion.

Site of treatment   Usually ECT is carried out in a special ECT suite. It may not be feasible to move a severely disturbed manic patient outside the ward and in this instance it may be preferable to carry out treatment in the patient's room if adequate resuscitation facilities are available.

Frequency of ECT   In the past, there has been a suggestion that more frequent ECT is necessary to treat severe disturbance. While there is certainly no place for multiple ECT in one day, treatment on consecutive days for up to three days may be beneficial. Following this, treatment on alternate days may be used. In some studies treatment was given three times per week with success. In general, manic patients will respond to a small number of treatments but in more severe cases between nine and twelve treatments may be required.

Concomitant medication   Clonazepam and carbamazepine are anticonvulsant and should be withdrawn if the patient fails to have adequate seizures. There have also been reports (Small *et al.* 1980) that the concomitant use of lithium during a course of ECT can lead to mild neurological and memory problems and may even interfere with the efficacy of the ECT.

Electrode placement   The introduction of increasingly sophisticated ECT technology has reopened the debate on the relative efficacy of unilateral versus bilateral treatment. Interestingly, a retrospective study (Small *et al.* 1985) indicated that unilateral ECT was not as effective as bilateral and, in some cases, actually worsened the patient's symptoms. It is therefore advisable to opt for bilateral ECT when treating mania.

Follow-up   ECT is likely to lead to good short-term response and often following a course of ECT the requirement for high-dose neuroleptics is considerably reduced. With improvement in mental state, oral medication may be accepted and prophylaxis with mood-stabilizing drugs should be started as soon as possible.

Clinical guidelines

Non-responders to neuroleptics
If sedation is required, barbiturates, clonazepam, or paraldehyde should be used. ECT is the treatment of choice and should be given bilaterally no less than three times weekly. As soon as compliance allows, oral mood-stabilizing drugs such as lithium or, if indicated, carbamazepine should be added.

Partial responders to neuroleptics
In the first instance lithium or carbamazepine should be added to the neuroleptic. Lithium is the drug of first choice but carbamazepine should

be used if the patient is a known lithium non-responder, has a history of rapid cycling, or there are medical or pharmacological contraindications to lithium use.

If a single drug addition fails then it is worth considering the combination of lithium and carbamazepine.

If the patient fails to respond to drug treatment or troublesome side-effects occur, then ECT should be considered.

References

Baldesarrini, R. and Stephens, J. (1970). Lithium carbonate for affective disorders. *Archives of General Psychiatry*, **22**, 72–5.

Baldesarrini, R. J., Cohen, B. M., and Teicher, M. (1988). Significance of neuroleptic dose and plasma level in the pharmacological treatment of psychosis. *Archives of General Psychiatry*, **45**, 79–91.

Black, D., Winokur, G., and Nasrallah, A. (1987). Treatment of mania: a naturalistic study of electroconvulsive therapy versus lithium in 438 patients. *Journal of Clinical Psychiatry*, **48**, 132–9.

Bouras, N., Traver, T., and Watson, U. P. (1982). Ward environment and disturbed behaviour. *Biological Medicine*, **12**, 309–19.

British National Formulary No.2 (1981). *4: Drugs acting on the central nervous system*, pp. 100–50. British Medical Association and The Pharmaceutical Society of Great Britain, London.

Chaudry, R. P. and Waters, B. G. H. (1983). Lithium and carbamazepine interaction: possible neurotoxicity. *Journal of Clinical Psychiatry*, **44**, 30–1.

Chouinard, G., Young, S. N., and Annable, L. (1983). Antimanic effect of clonazepam. *Biological Psychiatry*, **18**, 451–7.

Cohen, W. J. and Cohen, N. H. (1974). Lithium carbonate, haloperidol and irreversible brain damage. *Journal of the American Medical Association*, **230**, 1283–7.

Dubovsky, S. L., Franks, R. D., Allen, S., and Murphy, J. (1986). Calcium antagonists in mania: a double blind study of verapamil. *Psychiatry Research*, **18**, 309–20.

Dunner, D. L. and Fieve, R. R. (1974). Clinical factors in lithium prophylaxis failure. *Archives of General Psychiatry*, **30**, 229–33.

Fink, M. (1977). Mania and electro-seizure therapy. In *Manic illness* (ed. B. Shopsin), pp. 210–27. Raven Press, New York.

Folks, D. G., King, D., Dowdy, S. B., and Petrie, W. M. (1982). Carbamazepine treatment of selected affectively disordered inpatients. *American Journal of Psychiatry*, **143**, 115–17.

Giannini, A. J., Pascarzi, G. A., Loisell, R. H., Price, W. A., and Giannini, M. C. (1986). Comparison of clonidine and lithium in the treatment of mania. *American Journal of Psychiatry*, **143**, 1608–9.

Goodwin, F. K. and Zis, A. P. (1979). Lithium in the treatment of mania. *Archives of General Psychiatry*, **36**, 840–4.

Jefferson, J. W., Greist, J. H., Ackerman, D. L., and Carroll, J. A. (1987). *Lithium encyclopedia for clinical practice*, Second edition. American Psychiatric Press Inc., Washington DC.

Klein, E., Bental, E., Lerer, B., and Belmaker, R. (1984). Carbamazepine and haloperidol versus placebo and haloperidol in excited psychoses: a controlled study. *Archives of General Psychiatry*, **41**, 165–70.

Krauthammer, C. and Klerman, G. L. (1978). Secondary mania: manic syndromes associated with antecedent physical illness or drugs. *Archives of General Psychiatry*, **35**, 1333–9.

Lipinski, J. F. and Pope, H. G. (1982). Possible synergistic action between carbamazepine and lithium carbonate in the treatment of three acutely manic patients. *American Journal of Psychiatry*, **139**, 948–9.

Maggs, R. (1963). Treatment of manic illness with lithium carbonate. *British Journal of Psychiatry*, **109**, 56–65.

McCabe, M. (1976). ECT in the treatment of mania: a controlled study. *American Journal of Psychiatry*, **133**, 688–91.

McCabe, M. S. and Norris, B. (1977). ECT versus chlorpromazine in mania. *Biological Psychiatry*, **12**, 245–54.

Maguire, J. and Singh, A. N. (1987). Clonidine: an effective antimanic agent. *British Journal of Psychiatry*, **150**, 863–4.

Moss, G. and James, C. (1983). Carbamazepine and lithium carbonate synergism in mania. *Archives of General Psychiatry*, **40**, 588–9.

Peet, M. (1975). In *Lithium research and therapy* (ed. F.N. Johnson), pp. 25–41. Academic Press, London.

Post, R. M. (1982). Use of the anticonvulsant carbamazepine in primary and secondary affective illness: clinical and theoretical implications. *Psychological Medicine*, **12**, 701–4.

Post, R. M., Thomas, W., Uhde, M. D., and Putnam, F. W. (1982). Kindling and carbamazepine in affective illness. *Journal of Nervous and Mental Disease*, **170**, 717–31.

Schou, M. (1988). Lithium treatment: a refresher course. *British Journal of Psychiatry*, **149**, 541–7.

Shopsin, B., Gershon, S., Thompson, H., and Collins, P. (1975). Psychoactive drugs in mania. *Archives of General Psychiatry*, **32**, 34–42.

Small, J. G., Kellams, J. J., Milstein, V., and Small, I. F. (1980). Complications with electroconvulsive treatment combined with lithium. *Biological Psychiatry*, **15**, 103–13.

Small, J. G., Small, I. F., Milstein, V., Kellams, J. J., and Klapper, M. H. (1985). Manic symptoms: an indication for bilateral ECT. *Biological Psychiatry*, **20**, 125–34.

Small, J. G., *et al.* (1988). Electroconvulsive treatment compared with lithium in the management of manic states. *Archives of General Psychiatry*, **45**, 727–32.

Thomas, J. and Reddy, B. (1982). The treatment of mania: a retrospective evaluation of the effects of ECT, chlorpromazine and lithium. *Journal of Affective Disorders*, **4**, 85–9.

4

Anxiety disorders: what are the alternatives to prolonged use of benzodiazepines?

MICHAEL GELDER

There is, nowadays, broad agreement that anxiety disorders should not be treated by prolonged prescribing of benzodiazepine drugs because they can cause dependence. For this reason it has been proposed that these drugs should be taken for no more than four weeks (see, for example, Committee on Safety of Medicines 1988), but there has been no clear guidance about the treatment that should take the place of benzodiazepines in the treatment of chronic anxiety disorders, or indeed about the part these drugs should play in the management of acute cases. New kinds of anxiolytic drugs, antidepressants, and psychological treatments have all been proposed as alternatives to benzodiazepines, but there is no consensus about the indications for their use. Behind these disagreements concerning clinical practice lie more fundamental differences of opinion about the causes of chronic anxiety disorders, some believing that these are mainly 'biological', others that they are mainly psychological or social.

Controversies about the causes of anxiety disorders

There can be no doubt that both neurochemical and psychological processes are involved in normal anxiety. The neurochemical processes are concerned with neurotransmitters, their receptors, and second messengers, and it is reasonable to suggest that abnormalities of one or more of these processes could be the cause of anxiety disorders. The most fully studied of the neurotransmitter systems is the noradrenaline system which is involved in both the central control processes mediating anxiety and in peripheral effector mechanisms. Of the several other transmitter systems involved in anxiety, 5-hydroxytryptamine (5-HT) has received most attention recently with the development of new anxiolytic drugs acting on 5-HT systems. It follows from this view of anxiety disorders that treatment needs to correct the neurochemical disorder, and that drug treatment is the appropriate means of achieving this correction. Also, although a neurochemical theory of anxiety disorders allows a contribution from psychological factors, these are regarded

41

as secondary, accounting, for example, for the development of avoidance behaviour in people who have biologically determined panic attacks.

The alternative view of the aetiology of anxiety disorders is that the primary causes are psychological and social, and that any neurochemical changes are secondary. In this view, anxiety disorders occur when there are prolonged psychological or social problems, or when the person thinks or acts in a way that increases or prolongs anxiety. An important example of this kind of behaviour is persistent avoidance of situations that provoke anxiety, because this behaviour prevents the deconditioning and cognitive relearning which normally terminate an anxiety response. Another example is persistent low self-esteem, leading to anxiety in situations which other people would not find stressful. From this point of view, social or psychological treatment is required to deal with the fundamental disorder, and drug treatment can be only palliative, blocking the expression of anxiety but not affecting its basic causes.

The application of these ideas to the treatment of the various kinds of anxiety disorder will be considered in the later sections of this chapter. Before this is done, however, it is appropriate to consider the treatment of cases which are not sufficiently severe or chronic to meet diagnostic criteria for any of the anxiety disorders. These cases are common and their treatment is important because when it is carried out well many problems of dependence on anxiolytic drugs can be avoided.

Anxiety states of recent onset

The term anxiety state is used in the title of this section to include both anxiety disorders and cases which fail to meet the diagnostic criteria for any of the anxiety disorders. These are the conditions most often treated with benzodiazepine drugs, and since most improve quickly, prescribing can often be limited to the four weeks which has been recommended as the maximum period for the prescribing of these drugs without the risk of drug dependence (Committee on Safety of Medicines 1988).

Psychological treatments

Although these acute anxiety states can often be treated safely with a short course of a benzodiazepine, there is evidence that most cases respond equally well to simple psychological treatments. In one study conducted in general practice (Catalan *et al.* 1984), patients who would otherwise have been treated with benzodiazepines were allocated randomly either to benzodiazepine treatment or to brief counselling. The two groups had the same outcome, and the time needed for counselling (ten minutes per visit) was no greater than that given to the patients treated with drugs. An important finding of this study was that about a third of patients in both groups failed to improve and still met criteria for a 'case' at the six months follow-up assessments.

The results of a second study by the same group of investigators (Catalan and Gath 1990) suggest that one reason why this minority do not recover quickly is that the patients do not deal well with stressful circumstances. In this second study, patients resembling the poor prognosis group in the first investigation, were allocated randomly either to a more elaborate form of counselling or to drugs. In the counselling they were encouraged to learn better ways of solving problems; in the drug treatment the general practitioner selected the drug and dosage. At the end of treatment there were no significant differences between the two groups but at six months follow-up the problem-solving group had improved more than the drug-treated group and the outcome of the former resembled that of the good prognosis patients in the first trial. Since this counselling about problem solving requires rather more time and skill than the simple counselling used in the first study, a practical approach would be for the general practioner to begin with the simpler procedure, changing to problem solving (perhaps given by a counsellor attached to the practice) if the symptoms of anxiety do not subside quickly.

Medication

An alternative way of treating the more persistent of the anxiety states of recent onset is to use a non-benzodiazepine anxiolytic drug. The choice is between a beta-blocker, one of the new anxiolytic drugs that affect 5-HT, or an antidepressant drug. Beta-adrenergic blockers such as propranolol act rapidly but their therapeutic effects are limited to cases with prominent cardiovascular symptoms (see, for example, Hayes and Schulz 1987); and their effects of slowing the heart and precipitating asthma limit their use.

The new anxiolytic drugs affecting 5-HT systems are the azospirodecanediones, of which buspirone is an example. These compounds produce anxiolytic effects without sedation, and do not potentiate alcohol (Lader 1988). Despite these useful properties, their value in anxiety states of recent onset is limited because their anxiolytic effects take up to two weeks to be established (Uhlenhuth 1982), and usually a rapid therapeutic effect is needed. Because the drugs have not, so far, been shown to cause dependence, an obvious use would be in the treatment of anxiety states that do not resolve within the four weeks for which benzodiazepines can be used without risk of dependence. However, a high drop-out rate has been reported (45 per cent in the first month in the study of Rickels *et al.* 1988). Moreover, the strategy of starting with a benzodiazepine and changing to buspirone after three or four weeks may not be appropriate because there is some evidence that the effectiveness of buspirone is reduced after benzodiazepine treatment (Schweitzer *et al.* 1986; Cohn *et al.* 1986). Another possible use would be in the treatment of patients dependent on benzodiazepines but there is convincing evidence that buspirone does not prevent benzodiazepine withdrawal effects (Schweitzer and Rickels 1986; Lader and Olajide 1987).

Many patients with anxiety of recent onset have depressive as well as anxiety symptoms. The usual practice is to prescribe anxiolytics when anxiety symptoms predominate, and to prescribe an antidepressant when the depressive symptoms predominate. There is less agreement about the best treatment when the anxiety and depressive symptoms are of equal intensity: some use anxiolytics, others antidepressants. Since antidepressants have useful anxiolytic effects and do not cause dependence, it is generally better to prefer them to benzodiazepines for these cases. When compared with azospirodecanediones, the advantage of the antidepressant is the availability of drugs such as imipramine and amitriptyline which have been used for so many years that there is a confidence about their long-term safety which cannot yet be attained for the newer anxiolytics.

What then is the best treatment for anxiety states of recent onset? Although a benzodiazepine is a safe way of giving rapid relief of symptoms, simple counselling seems to be equally effective—and if limited to straight-forward support, no more time consuming. Benzodiazepines should not be used for more than four to six weeks because of the risk of dependence. When symptoms appear to be related to poor coping with life problems, general help in problem solving and assistance in dealing with specific problems can be offered by a suitably trained member of the general practice team. If further drug treatment is judged appropriate, a tricyclic antidepressant introduced while the benzodiazepine is reduced gradually, may be more appropriate than a drug of the azospirodecanedione series. If there are frequent panic attacks, or much phobic avoidance, early referral for the appropriate specialized psychological treatment should be considered.

The management of these cases of recent onset has been considered at length because it is in their treatment that anxiolytic drug dependence begins most often. However, whatever method of treatment is used for these early cases, some will still have symptoms six months later. The treatment of this important minority is the subject of the rest of this chapter, and generalized anxiety disorders will be considered first.

Generalized anxiety disorders

Until the recent concern about the prolonged use of benzodiazepine drugs, most cases of generalized anxiety disorder were treated in this way. Unfortunately, there is little firm evidence on which to base a choice between the available alternatives.

Medication

The important considerations about the drug treatment of generalized anxiety disorder are similar to those outlined above in relation to anxiety states of recent onset. Benzodiazepines relieve symptoms promptly but may cause dependence. Beta-blockers reduce cardiovascular symptoms of anxiety but do not have useful general therapeutic effects. The azospirode-canedione, buspirone, has been reported not to cause dependence when taken for six months (Rickels *et al.* 1989) but its anxiolytic effects take up to two weeks to appear; a high drop-out rate has been reported during its use for generalized anxiety disorder (Murphy *et al.* 1989) and, as noted above, it may not be fully effective when given after a benzodiazepine. Antidepressants are sometimes prescribed for generalized anxiety disorders, particularly when there are associated depressive symptoms, but there is no good evidence from clinical trials on which to judge their efficacy in these cases. Clinical experience suggests that the range of antidepressant dosage required to relieve anxiety is wider than that needed for an antidepressant effect: some patients respond to very small doses, others require the full dosage used to treat depressive disorders.

Psychological treatments

Of the psychological treatments for generalized anxiety disorders the simplest, relaxation training, has been in use for many years. Although often effective in minor anxiety problems, relaxation has not been shown to have a worthwhile therapeutic effect in generalized anxiety disorder (Le Boeuf and Lodge 1980), and there is no convincing evidence that yoga or meditation are of any more value in treating these disorders.

One reason for the poor results of relaxation with generalized anxiety disorder may be that it fails to deal with an important maintaining factor of these conditions, namely the worrying thoughts that generate and increase anxiety. Various kinds of cognitive therapy have been proposed to deal with these anxiety-provoking thoughts (see Barrios and Shigetomi 1979), and several studies have claimed good results. However, the studies showing beneficial effects of cognitive therapy have been concerned with minor anxiety problems rather than generalized anxiety disorders, and the few trials with the latter conditions have not shown substantial benefit (Woodward and Jones 1980; Jannoun *et al.* 1982).

The limited results of cognitive therapy in generalized anxiety disorder suggest that there may be another maintaining factor in addition to the anxious cognitions. It has been suggested that this factor may be avoidance behaviour, for although patients with generalized anxiety disorders do not engage in the obvious, focused avoidance behaviour that characterizes phobic patients, they avoid many diverse cues for anxiety (see Butler *et al.* 1987). Since avoidance is known to be a potent maintaining factor for anxiety disorders, treatments have been developed which combine cognitive therapy with exposure.

Two procedures of this kind have been developed, the first by Beck in the United States, the second by workers in Oxford. Beck has described promising results with anxiety disorders using an elaborate form of cognitive therapy (Beck *et al.* 1985), but the results of a clinical trial have not been published. In Oxford, Butler and coworkers (1987) have reported good results from two clinical trials of a simpler method which they call anxiety management (a combination of relaxation, exposure to avoided cues for anxiety, and simple cognitive procedures). Whether these procedures that include exposure lead to substantially better results than those of other psychological treatments for anxiety is still uncertain. Further clinical trials are required in which the several psychological treatments are compared with one another, and with anxiolytic drugs.

Although dynamic psychotherapy has been used for many years to treat generalized anxiety disorders, there are no satisfactory clinical trials on which to judge its efficacy. The principles of treatment can be summarized as focusing on ways of thinking that generate anxiety in situations that other people do not find threatening, and on the origin of these ways of thinking in past experience. Unlike cognitive therapy, such treatment includes the interpretation of thought content either in symbolic terms or as a reflection of experiences earlier in life. Clinical experience suggests that dynamic psychotherapy may help patients with chronic or recurrent anxiety symptoms resulting from difficulties in interpersonal relationships or intrapsychic problems clearly related to adverse experiences in childhood.

In the present state of knowledge it seems best to treat generalized anxiety disorders with anxiety management if this is available. If it is not, the prolonged use of benzodiazepines should be avoided by choosing another form of drug treatment. Given the present state of knowledge, an antidepressant seems preferable to prolonged use of buspirone because the long-term adverse effects have not yet been established with certainty. Because monoamine oxidase inhibitors (MAOIs) have more potential dangers than tricyclic antidepressants, and have not been shown to produce better results with patients with generalized anxiety disorders, a tricyclic should be the first line of treatment, with an MAOI used when tricyclics are ineffective. A minority of patients, whose recurrent anxiety symptoms are clearly related to intrapsychic problems arising from adverse childhood experiences, may be helped by dynamic psychotherapy.

Even if azospirodecanediones are eventually shown to be safe and effective anxiolytics, psychological treatment may still be better for chronic or recurrent cases because it may be effective not only for the current episode but also in preventing further episodes of anxiety. This preventive potential is there because patients are taught how to deal with the factors that can convert a short-lived anxiety response to stressful events into a prolonged anxiety disorder.

Panic disorder

Medication

The treatment of this disorder is particularly controversial. One view, developed originally by Klein and Fink (1962), is that panic disorder has biological causes and requires pharmacological treatment. Two kinds of drug treatment have been used for these cases: the first with the tricyclic antidepressant imipramine, the second with the triazolo-benzodiazepine, alprazolam (Sheehan 1987). The two drugs have been compared in a large multicentre clinical trial but the final results had not been published at the time of writing (May 1989). Preliminary reports of the investigation suggest that the therapeutic effects of the two drugs are similar, but that they differ in their unwanted effects. Unpleasant side-effects are more frequent in the early stages of treatment with imipramine, and withdrawal symptoms more frequent with alprazolam (Klerman 1988). The seriousness of the withdrawal symptoms after alprazolam is less when the drug dosage is reduced slowly but until more is known about this problem, a cautious approach is to use imipramine as the first line of drug treatment, reserving alprazolam for the minority who cannot tolerate the side-effects of imipramine. Before leaving the topic of drug treatment for panic disorder, it should be noted that although alprazolam is the benzodiazepine most often used for panic disorder, there is no convincing evidence that its therapeutic effects are different from those of comparable doses of other benzodiazepines.

Psychological treatments

The alternative to drug therapy is psychological treatment. This approach is based on the hypothesis that panic attacks develop when a person has a specific kind of fear of the sensations associated with autonomic arousal. The fear is that these sensations are an immediate forerunner of a serious medical emergency such as a heart attack or stroke. Although there is convincing evidence that patients with panic disorder have fears of this kind (see Clark *et al.* 1988), it is not yet certain whether these fears precede the panic attacks or are a consequence of experiencing the attacks. The truth will be difficult to establish, and in either case the fears could play a causative role: in the first case as a primary cause; in the second as a maintaining factor. Thus a more important consideration is whether changing anxious cognitions leads to improvement in panic disorder. There is growing evidence that a special form of cognitive therapy can achieve this effect (Clark *et al.* 1985) and is a practical form of treatment. However, this finding does not necessarily support the cognitive theory because two other psychological treatments have been reported to lead to improvement in panic disorder, and neither is specifically focused on changing the specific cognitions.

The first of these treatments is known as applied relaxation which is essentially a systematic training in rapid relaxation for use at times of anxiety. Using this form of treatment, Ost (1988) has reported short-term improvement in panic disorder of a magnitude comparable to that achieved with cognitive therapy. To date, long-term follow-up has not been reported.

Two other psychological treatments for panic disorder need to be considered. The first is an alternative form of cognitive behaviour therapy developed by Barlow (see Barlow 1988). This treatment differs from the cognitive therapy described by Clark and his collaborators in being less clearly focused on the specific fears of these patients, and using a wider range of anxiety management techniques. The reported results suggest that the effects of Barlow's treatment are similar to those of Clark's more focused form of cognitive treatment, but so far the two methods have not been compared directly. The second alternative is dynamic psychotherapy; this has not been evaluated in the treatment of panic disorder, but clinical experience does not suggest that it is effective.

Until the results of further clinical trials are available, the treatment of panic disorder will remain controversial. A practical approach is to use cognitive therapy or applied relaxation if an appropriately skilled therapist is available. Otherwise patients can be treated with drugs, starting with imipramine and using a benzodiazepine if the side-effects of imipramine cannot be tolerated. If imipramine is used, it should be started in very small doses and increased slowly to reduce drop-out rate; if alprazolam is used it should be withdrawn slowly to minimize withdrawal effects. Patients who do not respond to drug treatment or relapse after an initial response, can be referred for psychological treatment.

Agoraphobia

Until recently, most clinicians would have agreed that the behaviour therapy technique of exposure is the treatment of choice for agoraphobia unless there was a concurrent depressive disorder, in which case antidepressant drug therapy would be prescribed. Recently, an alternative treatment rationale has been suggested based on the idea (the 'biological hypothesis') that agoraphobia is not an independent entity but a variant of panic disorder. In this view (Klein *et al.* 1987) most cases of agoraphobia develop from panic disorder, through processes of conditioning and learning which lead to situational anxiety and avoidance behaviour. This hypothesis offers a plausible explanation for cases of agoraphobia with panic attacks but fails to account convincingly for the cases of agoraphobia that are not associated with panic attacks. Nor does it explain why many patients with established panic disorder have repeated panic attacks in situations which agoraphobics fear, and yet do not develop agoraphobia.

Medication

It follows from this 'biological' hypothesis that agoraphobia should be treated with imipramine or alprazolam to control the underlying panic disorder, and exposure for the situational anxiety and avoidance behaviour. If the biological hypothesis is correct, exposure and drug treatment should give better results than exposure alone; to date, studies addressing this question directly have concerned imipramine (rather than alprazolam). The results of these studies have been contradictory, some finding no benefit from the addition of imipramine to exposure (Marks *et al.* 1983; Cohen *et al.* 1984), others finding the combined treatment more effective (Mavissakalian *et al.* 1983; Telch *et al.* 1985). It has been suggested that the patients in these studies who benefited from imipramine had a depressive disorder as well as agoraphobia, but the published evidence does not support this explanation. It is more likely that the benefit of imipramine was due to its anxiolytic effect, which enhanced the effects of the weaker forms of exposure used in some studies, but not those of the stronger form of exposure treatment used in the study reporting no benefit from the addition of imipramine.

Psychological treatments

Individual and group forms of dynamic psychotherapy have been compared with exposure treatment as treatments for agoraphobia in a randomized clinical trial: both dynamic treatments were less effective than exposure (Gelder *et al.* 1967).

Despite many years of research into the treatment of agoraphobia, few of the crucially important comparisons between treatments have been made in convincing clinical trials. Until these studies have been performed, it seems best to treat agoraphobia with exposure treatment if an appropriately skilled therapist is available, using imipramine for patients who do not respond to exposure treatment, or are unwilling to take part in it—or when exposure treatment cannot be arranged. The evidence that there is benefit from combined exposure and imipramine is not sufficiently strong to recommend this procedure as a routine in the treatment of agoraphobia, but it is sufficient to justify a trial of combined treatment for patients who do not respond to exposure alone.

Social phobia

Patients with social phobia are often treated with anxiolytic drugs, but benzodiazepines should be avoided except for occasional use on important occasions. Two alternatives to benzodiazepines have been proposed: the use of other drugs, and cognitive-behavioural therapy.

Medication

Two drugs have been advocated: beta-blockers and monoamine-oxidase inhibitors. Although beta-blockers relieve the symptoms of musicians who experience anxiety when playing in public (James *et al.* 1974) and some kinds of specific performance anxiety, they do not produce similar benefit in most cases of social phobia (Gorman and Gorman 1987). The use of monoamine oxidase inhibitors for social phobia has been advocated recently, but to date (mid-1989) there has been no satisfactory clinical trial devoted specifically to these patients. The available evidence is from trials in which agoraphobic and social phobic patients were in the same treatment group, and it is impossible to reach definite conclusions about the response of the social phobics from the overall results.

Cognitive-behavioural treatment

This treatment for social phobia combines exposure to situations that have been avoided, and anxiety management techniques to control anxiety experienced in the phobic situations or when anticipating them. For these patients, the technique of exposure has to be a little different from that used for other phobic disorders. The change is necessary because social phobics seldom avoid situations completely; more often they enter the situations while avoiding some particular aspect such as making eye contact or starting a conversation. The combination of anxiety management and an appropriate kind of exposure results in more long-term improvement than exposure alone (Butler *et al.* 1984). As yet, this kind of cognitive-behavioural therapy has not been compared directly with MAOIs so that it is difficult to give firm advice about the best kind of treatment for social phobia. Nevertheless, the high relapse rate reported when MAOIs are withdrawn, and the good-long term results reported with cognitive-behavioural treatment, suggest that the latter may be the better treatment.

There has been no satisfactory clinical trial of dynamic psychotherapy for social phobia. Clinical experience suggests that this form of treatment is less effective than cognitive-behavioural therapy when social anxiety is clearly limited to specific situations. However, dynamic psychotherapy does seem to be valuable when there are widespread social anxieties arising from intrapsychic problems such as low self-esteem, and these problems are clearly related to adverse experiences in childhood.

Clinical guidelines

1. The controversy about the long-term use of benzodiazepines is now resolving in favour of a general view that although these drugs are valuable for short-term use, they should not be continued for more than four to six weeks. The possible exception to this general rule is in the

treatment of panic disorder for which long-term treatment with alprazolam has been claimed to produce good results. There is much less agreement about the best alternative to benzodiazepines. For each type of anxiety disorder the main choice is between a non-benzodiazepine anxiolytic drug and a form of cognitive-behavioural therapy. At present, the choice of drug is between a tricyclic antidepressant, with a MAOI as second choice, and one of the azospirodecanediones such as buspirone. Until more information is available about the side-effects, toxicity, and potential for dependence of these new drugs, a conservative view is that they should be used only for periods comparable to those recommended for benzodiazepines—a recommendation that would limit their value in clinical practice.

2. Cognitive-behavioural treatments work best when they combine relaxation, exposure, and measures to control 'fear of fear'. The emphasis on each of these techniques varies according to the type of anxiety disorder. For panic disorder it seems important to focus treatment on the specific fears that characterize this condition. For phobic disorders a combination of exposure and more general anxiety management techniques is effective. For generalized anxiety disorders a broad approach seems to be needed, combining relaxation, cognitive procedures, and exposure.

3. Dynamic psychotherapy has not been tested systematically as a treatment for anxiety disorders. It appears to be unnecessary for disorders of recent onset and less effective than cognitive-behavioural therapy for established disorders with phobic or panic symptoms. Clinical experience suggests that dynamic psychotherapy is of value in treating patients whose longstanding anxiety symptoms are clearly related to serious intrapsychic problems arising from experiences in childhood and early adolescence.

In all areas of medicine, arguments about clinical practice flourish when facts are few. The controversies about the treatment of chronic anxiety disorders will be settled only when more clinical trials have been carried out and, more particularly, when research on drug and psychological treatments are integrated more closely than they have been so far.

References

Barlow, D. H. (1988). *Anxiety and its disorders: the nature and treatment of anxiety and panic*. Guilford Press, New York.

Barrios, B. A. and Shigetomi, C. C. (1979). Coping-skills training for the management of anxiety: a critical view. *Behaviour Therapy*, **10**, 491–522.

Beck, A. T., Emery, G., and Greenberg, R. (1985). *Anxiety disorders and phobias: a cognitive perspective*. Basic Books, New York.

Butler, G., Cullington, A., Munby, M., Amies, P., and Gelder, M. G. (1984).

Exposure and anxiety management in the treatment of social phobia. *Journal of Consulting and Clinical Psychology*, **52**, 642–50.

Butler, G., Gelder, M. G., Hibbert, G., Cullington, A., Klimes, L. (1987). Anxiety management—developing effective strategies. *Behaviour Research and Therapy*, **25**, 517–22.

Catalan, J. and Gath, D. H. (1989). (personal communication)

Catalan, J., Gath, D. H., Edmonds, G., and Ennis, J. (1984). The effects of non-prescribing of anxiolytics in general practice: I, controlled evaluation. *British Journal of Psychiatry*, **144**, 593–602.

Clark, D. M., Salkovskis, P. M., and Chalkley, A. J. (1985). Respiratory control as a treatment for panic attacks. *Journal of Behaviour Therapy and Experimental Psychiatry*, **16**, 23–30.

Clark, D. M., *et al.* (1988). Tests of a cognitive theory of panic. In *Panic and Phobias: 2.* (ed. I. Hand and H. V. Wittchen) pp. 149–58. Springer-Verlag, Berlin.

Cohen, S. D., Monteiro, W., and Marks, I. M. (1984). Two-year follow-up of agoraphobics after exposure and imipramine. *British Journal of Psychiatry*, **144**, 276–81.

Cohn, J. B., Bowden, C. L., Fisher, J. G., and Rodos, J. J. (1986). Double-blind comparison of buspirone and clorazepate in anxious out-patients. *American Journal of Medicine*, **80**, (suppl. 36), 10–16.

Committee on Safety of Medicines (1988). Benzodiazepines, dependence and withdrawal symptoms. *Current Problems*, **21**, 1–2.

Gelder, M. G., Marks, I. M., Wolff, H. H. and Clarke, M. (1967). Desensitization and psychotherapy in the treatment of phobic states: a controlled inquiry. *British Journal of Psychiatry*, **113**, 53–73.

Gorman, J. M. and Gorman, L. K. (1987). Drug treatment of social phobia. *Journal of Affective Disorders*, **13**, 183–92.

Hayes, P. E. and Schulz, S. C. (1987). Beta-blockers in anxiety disorders. *Journal of Affective Disorders*, **13**, 119–30.

James, I. M., Griffith, D. N. W., Pearson, R. M., and Newby, P. (1977). Effect of oxprenolol on stage-fright in musicians. *Lancet*, **ii**, 952–4.

Jannoun, L., Oppenheimer, C., and Gelder, M. G. (1982). A self-help treatment programme for anxiety state patients. *Behaviour Therapy*, **13**, 103–11.

Klein, D. F. and Fink, M. (1962). Psychiatric reaction patterns to imipramine. *American Journal of Psychiatry*, **119**, 432–8.

Klein, D. F., Ross, D. C., and Cohen, P. (1987). Panic and avoidance in agoraphobia: application of path analysis to treatment studies. *Archives of General Psychiatry*, **44**, 377–85.

Klerman, G. L. (1988). Overview of the cross-national collaborative panic study. *Archives of General Psychiatry*, **45**, 407–12.

Le Boeuf, A. and Lodge, J. A. (1980). A comparison of frontalis EMG feedback training and progressive relaxation in the treatment of chronic anxiety. *British Journal of Psychiatry*, **137**, 279–84.

Lader, M. and Olajide, D. (1987). A comparison of buspirone and placebo in relieving benzodiazepine withdrawal symptoms. *Journal of Clinical Psychopharmacology*, **7**, 11–15.

Lader, M. (1988). The practical use of buspirone. In *Buspirone: a new introduction to the treatment of anxiety*. (ed. M. Lader), pp. 21–7. Royal Society of Medicine International Congress and Symposium Series no. 133.

Marks, I. M., Gray, S., Cohen, D., Hill, R., Mawson, D., Ramm, E., and Stern, R.S. (1983). Imipramine and brief therapist-aided exposure in agoraphobics having self-exposure homework. *Archives of General Psychiatry*, **40**, 153–62.

Mavissakalian, M., Michelson, L., and Dealy, R. S. (1983). Pharmacological treatment of agoraphobia: imipramine versus imipramine with programmed practice. *British Journal of Psychiatry*, **143**, 348–55.

Murphy, S. M., Owen, R., and Tyrer, P. (1989). Comparative assessment of efficacy and withdrawal symptoms after 6 and 12 week treatment with diazepam or buspirone. *British Journal of Psychiatry*, **154**, 529–34.

Ost, L. G. (1988). Applied relaxation vs progressive relaxation in the treatment of panic disorder. *Behaviour Research and Therapy*, **26**, 13–23.

Rickels, K., Schweizer, E., Cscnalosi, I., Case, W. G., and Chung, H. (1988). Long-term treatment of anxiety and risk of withdrawal: prospective comparison of clorazepate and buspirone. *Archives of General Psychiatry*, **45**, 444–50.

Rickels, K., Schweizer, E., and Case, W. G. (1989). Withdrawal problems with anti-anxiety drugs: nature and management. In *Psychopharmacology of anxiety*. (ed. P. Tyrer), pp. 283–93). Oxford University Press.

Schweitzer, E. and Rickels, K. (1966). Failure of buspirone to manage benzodiazepine withdrawal. *American Journal of Psychiatry*, **143**, 1590–2.

Schweitzer, E., Rickels, K., and Lucki, I. (1986). Resistance to the anti-anxiety effect of buspirone in patients with a history of benzodiazepine use. *New England Journal of Medicine*, **314**, 719–20.

Sheehan, D. V. (1987). Benzodiazepines in panic disorder and agoraphobia. *Journal of Affective Disorders*, **13**, 169–81.

Telch, M. J., Agras, W. S., Taylor, W. B., Roth, W. T., and Gallen, C. C. (1985). Combined pharmacological and behavioural treatment of agoraphobia. *Behaviour Research and Therapy*, **23**, 325–35.

Uhlenhuth, E. H. (1982). Buspirone: a clinical review of a new non-benzodiazepine anxiolytic. *Journal of Clinical Psychiatry*, **43**, 109–14.

Woodward, R. and Jones, R. B. (1980). Cognitive restructuring treatment: a controlled trial with anxious patients. *Behaviour Research and Therapy*, **18**, 401–7.

5

Cognitive therapy for depression and anxiety: is it better than drug treatment in the long term?

DAVID M. CLARK

Cognitive therapy aims to teach patients to identity and modify both the negative thoughts and behaviours which, according to cognitive theories, maintain anxiety and depression, and the dysfunctional beliefs which it is suggested make some people prone to becoming depressed or anxious. If successful in achieving these aims, one might suppose that cognitive therapy would be more effective than drug treatments in the long term. This chapter reviews the literature relevant to this question and concludes that there is tentative support for the long-term superiority of cognitive therapy but that there is a need for considerably more research, particularly with some of the anxiety disorders.

Depression

Several studies have shown that cognitive therapy is at least as effective as tricyclic antidepressants in the acute treatment of non-psychotic, non-bipolar depressed out-patients. Encouraging as it is, this finding would have restricted clinical utility unless it could also be shown either that individuals who respond to cognitive therapy are different from those who respond to tricyclic antidepressants and/or that cognitive therapy is associated with a lower relapse rate. Several studies have investigated the question of differential relapse and each provides some support for the superiority of cognitive therapy. However, before we discuss these studies it is important to clarify what is meant by relapse.

Defining relapse

Relapse is typically defined either in terms of return to treatment or of exceeding a clinical cut-off on a symptom measure. Neither criterion is entirely satisfactory. It has been argued that the return to treatment measure may favour cognitive therapy, as patients who relapse after taking antidepressants may think that the best way of dealing with the relapse is to seek further antidepressant treatment. Patients who relapse after cognitive therapy, however may initially attempt to deal with the relapse by not seeking treatment but rather by using the skills they have learnt in therapy. Symptomatic measures may underestimate relapse because patients who believe they are beginning to relapse may seek early treatment, respond to it, and then never reach the symptom criterion. In view of these problems, one has greatest confidence in findings which hold up using both measures.

Klerman (1978) has suggested a useful distinction between symptomatic *relapse*, the return of symptoms associated with the prior episode, and *recurrence*, the onset of symptoms associated with a wholly new episode. As it has been suggested that the average duration of an untreated episode of depression is approximately nine months, follow-ups covering the first six months after a three-month course of cognitive therapy will mainly detect relapse while longer follow-ups will detect a mixture of both relapse and recurrence. At present the longest follow-up is two years, which is probably too short to pick up the majority of recurrences as it has been estimated (Evans *et al.* 1989) that the typical out-patient depressive can be expected to have a new episode approximately every three years.

The studies

Four studies have compared the long-term outcome of cognitive therapy and pharmacotherapy. In all the studies cognitive therapy was given for three months while the duration of antidepressant medication varied from 3–15 months. Table 2 summarizes the design of each study. Kovacs and coworkers (1981) used both return-to-treatment and symptomatic relapse criteria and followed up patients for 12 months after the end of cognitive therapy. Simmons *et al.* (1986) and Blackburn *et al.* (1986) mainly used the return-to-treatment criterion and followed up patients for respectively 12 and 24 months after the end of cognitive therapy. Evans *et al.* (1989) used both return-to-treatment and symptomatic criteria and obtained essentially similar results with both measures. They followed up patients for a total of 24 months after the end of cognitive therapy. As it is well established that continuing pharmacotherapy after recovery is associated with a lower relapse rate (Prien and Kupfer 1986), comparisons relating to pharmacotherapy with and without maintenance will be discussed separately.

Cognitive therapy versus pharmacotherapy without maintenance medication

Three studies have compared cognitive therapy with pharmocotherapy without maintenance. The relapse figures for each study are given in Table 2. In each study, tricyclics were withdrawn after approximately three months. Simmons *et al.* (1986) and Evans *et al.* (1989) found that there was no difference in initial response but patients successfully treated with cognitive therapy were significantly less likely to relapse. Kovacs *et al.* (1981) also found some evidence of differential long-term outcome. Pharmacotherapy patients had significantly higher Beck Depression Inventory

Table 2 *Summary of depression follow-up studies*

Study	Comparison groups and initial outcome	Length of follow-up (months)	Who followed up?	Relapse criterion	Percentage of patients relapsing in follow-up period				
					C.T.	TCA (no maintenance)	TCA (6 months maintenance)	TCA (12 months maintenance)	CT plus TCA
Kovacs *et al.* 1981	CT > TCA	12	Treatment completers, irrespective of initial outcome	Symptom (BDI ≥16) and/or return to treatment	33%	59%	—	—	—
Simmons *et al.* 1986	CT=TCA =CT+PL =CT+TCA	12	Treatment responders (BDI ≤9)	Symptom (BDI≥16) and/ or return to treatment	20%	66%	—	—	43%
Blackburn *et al.* 1986	*Hospital sample* CT=TCA CT+TCA>CT CT+TCA>TCA *G.P. sample* CT>TCA CT+TCA=CT CT+TCA>TCA	24	Treatment responders (BDI ≤8)	Return to treatment	23%	—	78%	—	21%
Evans *et al.* 1989	CT=TCA CT+TCA>TCA CT+TCA=CT	24	Treatment responders (BDI ≤15)	Symptom (2 BDIs ≥16 separated by at least seven days)	20%	50%	—	27%	15%

Note: CT=Cognitive therapy; TCA=Tricyclic antidepressant; PL=Placebo

(BDI) scores at one year follow-up and were twice as likely to relapse. However, the latter comparison was not significant. Unlike these studies by Simmons *et al.*, and Evans *et al.*, the study of Kovacs *et al.* had a differential attrition rate, with more patients dropping out of pharmacotherapy than cognitive therapy. As the follow-up phase was restricted to treatment completers, Kovacs and coworkers may therefore have under-estimated the difference in long-term outcome between the two treatments.

Cognitive therapy versus pharmacotherapy with maintenance

It is currently considered good clinical practice to maintain patients on tricyclics for at least four months after recovery (Prien and Kupfer 1986). Blackburn *et al.* (1986) compared patients who had responded to three months of cognitive therapy with patients who had responded to three months of pharmacotherapy and then received a further six months of maintenance treatment. Cognitive therapy was again associated with a significantly lower relapse rate.

Finally, one study (Evans *et al.* 1989) included a comparison between patients who initially responded to cognitive therapy and patients who responded to tricyclics within three months and were then maintained on medication for a further 12 months. Cognitive therapy and long-term maintenance pharmacotherapy showed similar relapse rates and these were significantly lower than pharmacotherapy without maintenance. However, it should be noted that patients given maintenance pharmacotherapy had only been followed-up for one year after the end of treatment while those given cognitive therapy had been followed up for two years. It may be that as the follow-up period in this study increases, and the difference between conditions in the time since the end of treatment decreases, maintenance pharmacotherapy will appear relatively less effective. This would be consistent with the findings of work investigating predictors of relapse. Evans *et al.* (1989) found that the only biochemical or psychological variable which predicted relapse over and above residual depression was a measure of negative attributional style (tendency to attribute negative life events to personal inadequacy), one of the cognitive characteristics which is said to make individuals prone to becoming depressed (Abramson *et al.* 1978; Peterson and Seligman 1984). Patients who responded to pharmacotherapy showed a significantly smaller drop in negative attributional style than patients who responded to cognitive therapy, suggesting that in the long term they should be more vulnerable to a recurrence.

Combined therapy

So far we have concentrated on comparisons between cognitive therapy alone and pharmacotherapy alone. Four studies (Murphy *et al.* 1984; Hollon *et al.* 1989; Blackburn *et al.* 1986; Beck *et al.* 1985) have also included a group given a combination of cognitive therapy and pharmacotherapy (with no maintenance). These studies failed to find any consistent differences in either initial response or relapse between cognitive therapy alone and cognitive therapy plus medication. There is therefore no convincing empirical support for the idea that the relatively good long-term outcome observed with cognitive therapy can be significantly enhanced by the concurrent use of medication.

In conclusion, the above studies suggest that in the immediate long term (one to two years), well-conducted cognitive therapy is associated with better outcome than pharmacotherapy without maintenance or pharmacotherapy with six months maintenance, but cognitive therapy may not be superior to pharmacotherapy with 12 months maintainance. However, two words of caution are in order. First, the apparent superiority of cognitive therapy over pharmacotherapy with six months maintenance is based on the results of a single study which used the return-to-treatment criterion, a criterion which, it has been suggested, may unfairly favour cognitive therapy. Secondly, so far we have tacitly assumed that the same individuals are likely to show a good response to both cognitive therapy and pharmacotherapy. However, this is not entirely correct. Simmons *et al.* (1985) found that although approximately equal numbers of patients showed a good initial response to cognitive therapy and pharmacotherapy, those patients who scored high on a measure of coping skills (learned resourcefulness) at the start of treatment responded relatively better to cognitive therapy, while those who scored low on learned resourcefulness responded relatively better to pharmacotherapy. This suggests that in some individuals (with high learned resourcefulness) the apparent long-term superiority of cognitive therapy may have been underestimated while in others (with low learned resourcefulness) it may have been overestimated.

Anxiety disorders

Cognitive therapy is a relatively recent development in the treatment of anxiety disorders. Following the successful application of cognitive therapy in depression, Beck and other investigators have turned their attention to developing forms of cognitive treatment for anxiety. Most of these treatments have built on existing behavioural techniques and are probably best described as forms of cognitive behaviour therapy. To date there are no studies directly comparing the long-term effects of cognitive behaviour therapy and drug treatment in the anxiety disorders. In the absence of such studies, any conclusions that we draw are necessarily preliminary and must be based on comparisons between studies which themselves investigate only one of the two treatments. Such studies are themselves relatively rare, particularly in the case of drug treatments where the majority of reports include only end-of-treatment data. Our general conclusion will be that, when it has been investigated, cognitive-behavioural treatments appear to have good long-term outcome with high initial response rates and treatment gains maintained at follow-ups ranging between six months and several years. By contrast in several anxiety disorders there are no compelling reasons for choosing drug treatments, and in those where drugs have been associated with good initial response (for example, alprazolam and imipramine in panic disorder) preliminary studies show high relapse rates after discontinuation of the drug, suggesting that the long-term value of the drugs is severely limited.

Panic disorder

In recent years considerable progress has been made in developing an effective form of cognitive therapy for panic. This treatment concentrates on helping patients to identify and modify the catastrophic interpretations of bodily sensations which cognitive theories (e.g. Clark 1986) suggest play a causal role in the production of attacks. Consecutive case series and one controlled trial (see Rachman and Maser (1988) for a review) indicate that between 80 per cent and 100 per cent of patients are panic free at the end of treatment. In the one study (Clark *et al*. 1985) which investigated long-term follow-up, these gains were maintained over the next two years. Two drugs—the triazolo-benzodiazepine, alprazolam, and the tricyclic antidepressant, imipramine—have frequently been advocated as treatments for panic disorder. A recent multi-centre clinical trial found that in the short term (eight weeks) both drugs were significantly more effective than placebo. No long-term (one year or greater) follow-up data is available. However, studies of relapse rates during the two to three months after drug withdrawal suggest that both drugs will probably be associated with poor long-term outcome. For example, Sheehan (1986) reported that between 70 per cent and 90 per cent of his patients experienced a relapse in the three months immediately after withdrawal from alprazolam, phenelzine, or imipramine. Similarly, Pecknold and co-workers (1988) found that two weeks after withdrawal from alprazolam, patients were experiencing as many panic attacks as they had before the start of treatment. Unless these deteriorations are transient phenomena, the long-term outcome of pharmacotherapy is likely to be poor.

Agoraphobia

Cognitive therapy for agoraphobia is mainly based on the hypothesis that agoraphobia is not an independent entity but a variant of panic disorder and treatment progresses in a similar way to the cognitive treatment of panic disorder but with a stronger emphasis on the behaviour therapy technique of *in vivo* exposure to feared situations. Consecutive case series again suggest a good initial response which is maintained at long-term (two years) follow-up (Clark *et al*. 1985). A more extensive literature has investigated the long-term outcome of *in vivo* exposure alone. Approximately 60–70 per cent of patients showed clinically significant improvements immediately after treatment (Jansson and Ost 1982) and these are largely maintained at follow-ups ranging from six months to eight years (Burns *et al.* 1986). However, temporary stress-induced setbacks occur and a significant proportion of patients seek further treatment during the follow-up period.

The most consistently advocated drug treatment for agoraphobia is imipramine. In general imipramine is combined with *in vivo* exposure and there is evidence that *some* patients show better initial response to the combined treatment than to exposure alone (see Telch (1988) for a review). However, continued superiority of the combined approach over exposure alone has not been demonstrated after patients have been withdrawn from medication.

Social phobia

Recent studies (Butler *et al.* 1984; Mattick and Peters 1988) have shown that cognitive behaviour therapy is an effective treatment for social phobia in the short term and that improvements are either maintained or increased at up to six months follow-up. In the Butler *et al.* (1984) study, return to treatment (but not symptomatology) was also assessed at 12 months. None of the patients given cognitive behaviour therapy requested further treatment.

To date, no drugs have been shown to be specifically effective in the treatment of social phobia though it has recently been claimed that monoamine oxidase inhibitors may be effective.

Generalized anxiety disorder

There are no long-term follow-ups of treatment for generalized anxiety. However, Butler *et al.* (1987) have recently shown that cognitive behaviour therapy is more effective than no treatment and that the gains obtained are maintained at six months follow-up though several patients sought further treatment during this period. For a long time, benzodiazepines were a popular treatment for generalized anxiety but increasing awareness of dependence problems has led to the recommendation that these drugs should be taken for no longer than six weeks. Following this recommendation, Lindsay *et al.* (1987) compared patients who were given lorazepam for a total of four weeks with patients who received one of two different types of cognitive-behavioural treatment. Lorazepam was associated with marked initial improvement which declined as the trial progressed and was minimal at the end of therapy. In contrast, patients given cognitive-behavioural treatment continued to improve throughout treatment and this improvement was maintained at three months follow-up.

General conclusions and recommendations

Depression

Cognitive therapy and tricyclic antidepressants are well validated short-term treatments for depression. Approximately 60 per cent of patients become asymptomatic after three months of either treatment. Tapering medication at this stage (and perhaps also up to six months later) is associated with a high relapse rate during the next one to two years. Significantly lower, and comparable, relapse rates are achieved with either short-term cognitive therapy or long-term (12 months) maintenance tricyclics. As individuals who respond best to, or find acceptable, cognitive therapy are not entirely the same as those who respond best to, or find acceptable, maintenance tricyclics, both treatments should have an important place in the management of depression.

Although cognitive therapy and long-term maintenance tricyclics are significant advances in the treatment of depression, both are considerably less effective than one would wish. In the Evans *et al.* (1989) study, for example, only 38 per cent of the patients who were offered cognitive therapy found the treatment acceptable, responded to it in three months, and stayed well for the next two years. The comparable figure for maintenance tricyclics is 31 per cent.

Anxiety disorders

For many of the anxiety disorders there are no well-validated pharmacological treatments, and cognitive-behavioural interventions are the treatment of choice. However, in those disorders which are characterized by recurrent panic attacks, alprazolam and imipramine are effective in the short term. No long-term follow-ups exist but data on relapse immediately following discontinuation of the drugs suggests that neither are likely to be effective in the long term. This is in marked contrast to the preliminary data available on cognitive-behavioural treatments which suggests that the substantial initial gains obtained with these treatments are maintained over the next two years at least. Given these findings, it would seem that the treatment of choice for agoraphobia and panic disorder is cognitive behaviour therapy. If patients fail to respond, tricyclic antidepressants might be considered as an adjunct, particularly if the patient is also depressed.

The following books are recommended for clinicians who wish to acquaint themselves with the practical details of how to do cognitive therapy: Beck, A. T., Rush, A. J., Shaw, B. F., and Emery, G. (1979). *Cognitive therapy of depression*. Guilford Press, New York; Beck, A. T., Emery, G. and Greenberg, R. L. (1985). *Anxiety Disorders and phobias: a cognitive perspective*. Basic Books, New York; Blackburn, I. M. (1987). *Coping with depression*. Chambers, Edinburgh; Burns, D. D. (1980). *Feeling good*. New American Library, New York; Hawton, K., Salkovskis, P. M., Kirk, J. and Clark, D. M. (1989). *Cognitive behaviour therapy for psychiatric problems: a practical guide*. Oxford University Press; Williams, J. M. G. (1984) *The psychological treatment of depression: a guide to the theory and practice of cognitive behaviour therapy*. Croom Helm, London.

References

Abramson, L. Y., Seligman, M. E. P., and Teasdale, J. D. (1978). Learned helplessness in humans: critique and reformulation. *Journal of Abnormal Psychology*, **87**, 49–74.

Beck, A. T., Hollon, S. D., Young, J. E., Bedrosian, R. C., and Budenz, D. (1985). Treatment of depression with cognitive therapy and amitriptyline. *Archives of General Psychiatry*, **42**, 142–8.

Blackburn, I. M., Eunson, K. M., and Bishop, S. (1986). A two-year naturalistic follow-up of depressed patients treated with cognitive therapy, pharmacotherapy, and a combination of both. *Journal of Affective Disorders*, **10**, 67–75.

Burns, L. E., Thorpe, G. L., and Cavallara, L. A. (1986). Agoraphobia 8 years after behavioral treatment: a follow-up study with interview, self-report and behavioral data. *Behavior Therapy*, **17**, 580–91.

Butler, G., Cullington, A., Munby, M., Amies, P., and Gelder, M. (1984). Exposure and anxiety management in the treatment of social phobia. *Journal of Consulting and Clinical Psychology*, **52**, 642–50.

Butler, G., Cullington, A., Hibbert, G., Klimes, I., and Gelder, M. (1987). Anxiety management for persistent generalised anxiety. *British Journal of Psychiatry*, **151**, 535–42.

Clark, D. M. (1986). A cognitive approach to panic. *Behaviour Research and Therapy*, **24**, 461–70.

Clark, D. M., Salkovskis, P. M., and Chalkey, J. (1985). Respiratory control as a treatment for panic attacks. *Journal of Behavior Therapy and Experimental Psychiatry*, **16**, 23–30.

Evans, M. D., Hollon, S. D., DeRubeis, R. J., Piasecki, J. M., Grove, W. M., Garvey, M. J., and Tuason, V. B. (1989). Differential relapse following cognitive therapy, pharmacotherapy, and combined cognitive-pharmacotherapy for depression: IV. A two-year follow-up of the CPT project. Manuscript submitted for publication.

Hollon, S. D., DeRubeis, R. J., Evans, M. D., Weimer, M. J., Garvey, M. T., Grove, W. M., and Tuason, V. B. (1989) Cognitive therapy, pharmacotherapy, and combined cognitive-pharmacotherapy in the treatment of depression: I. Differential outcome of the CPT project. Manuscript submitted for publication.

Jansson, L. and Ost, L. G. (1982), Behavioural treatments for agoraphobia: an evaluative review. *Clinical Psychology Review*, **2**, 311–36.

Klerman, G. L. (1978). Long-term treatment of affective disorders. In *Psychopharmacology: a generation of progress*, (ed. M. A. Lipton, A. DiMascio, and K. Killam), pp. 1302–11. Raven Press, New York.

Kovacs, M., Rush, A. J., Beck, A. T., and Hollon, S. D. (1981). Depressed outpatients treated with cognitive therapy or pharmacotherapy: one-year follow-up. *Archives of General Psychiatry*, **38**, 33–9.

Lindsay, W. R., Gamsu, C. V., McLaughlin, E., Hood, E. M., and Elspie, C. A. (1987). A controlled trial of treatments for generalized anxiety. *British Journal of Clinical Psychology*, **26**, 3–16.

Mattick, R. P. and Peters, L. (1988). Treatment of severe social phobia: effects of guided exposure with and without cognitive restructuring. *Journal of Consulting and Clinical Psychology*, **56**, 251–60.

Murphy, G. E., Simmons, A. D., Wetzel, R. D. and Lustman, P. J. (1984). Cognitive therapy and pharmacotherapy: singly and together in the treatment of depression. *Archives of General Psychiatry*, **41**, 33–41.

Pecknold, J. C., Swinson, R. P., Kuch, K. and Lewis, C. P. (1988). Alprazolam in panic disorder and agoraphobia: results from a multicenter trial. III. Discontinuation effects. *Archives of General Psychiatry*, **45**, 429–36.

Peterson, C. and Seligman, M. E. P. (1984). Causal explanations as a risk factor for depression: theory and evidence. *Psychological Review*, **91**, 347–74.

Prien, R. F. and Kupfer, D. J. (1986). Continuation drug therapy for major depressive episodes: how long should it be maintained? *American Journal of Psychiatry*, **143**, 18–23.

Rachman, S. and Maser, J. D. (1988). *Panic: psychological perspectives*, Lawrence Erlbaum, Hillsdale NJ.

Sheehan, D. V. (1986). Tricyclic anti-depressants in the treatment of panic and anxiety disorders. *Psychosomatics*, **27**, 10–16.

Simmons, A. D., Lustman, P. J., Wetzel, R. D., and Murphy, G. E. (1985). Predicting response to cognitive therapy of depression: the role of learned resourcefulness. *Cognitive Therapy and Research*, **9**, 79–89.

Simmons, A. D., Murphy, G. E., Levine, J. L., and Wetzel, R. D. (1986). Cognitive therapy and pharmacotherapy for depression: sustained improvement over one year. *Archives of General Psychiatry*, **43**, 43–8.

Telch, M. J. (1988). Combined pharmacological and psychological treatments for panic sufferers. In *Panic: psychological perspectives*, (ed. S. Rachman and J. D. Maser), pp. 167–88. Lawrence Erlbaum, Hillsdale NJ.

6

The suicidal patient: how should the therapist respond?

MARJORIE E. WEISHAAR and AARON T. BECK

Introduction

Responding therapeutically to suicidal patients requires considerable energy and creativity. Managing a suicidal crisis can generate anxiety for even the most experienced clinician. This chapter presents a description of how cognitive therapy provides a structured approach to both acute suicidal episodes and to the treatment of more chronic cognitive deficits which may predispose the patient to future suicidal behaviour.

While we are describing the treatment of suicidal thoughts and behaviour from the perspective of cognitive therapists, this approach can be incorporated into the work of general psychiatrists who treat suicidal individuals. The role of hospitalization will not be considered here. Hospitalization is recommended in cases in which the patient cannot otherwise be safeguarded, particularly when intense and continuous suicidal ideation accompanies an affective disorder or psychosis.

Research in cognitive therapy has focused on depressed suicidal individuals. Cognitive therapy has demonstrated efficacy in the treatment of unipolar depression (see Chapter 5), with both out-patients and hospitalized patients. Thus these techniques may also be used while the suicidal person is hospitalized.

The cognitive model

Cognitive therapy, developed from empirical studies of depressed patients, is based on a model of psychopathology which maintains that how a person perceives and interprets the environment largely determines how he or she feels and behaves. Depressed patients were found to have a negative bias to their thinking, perceiving themselves as inadequate, the world as harsh and punishing, and the future as bleak and hopeless. (For a comprehensive guide to cognitive therapy for depression, the reader is referred to Beck *et al.* 1979*b*.) This negative bias in thinking is maintained by systematic errors in logic, called cognitive distortions. These include:

(1) *dichotomous thinking*, which categorizes experience in absolute categories (i.e., black and white thinking);

(2) *arbitrary inference*, or drawing a conclusion in the absence of supportive evidence;

(3) *selective extraction*, which entails attending to only a portion of relevant data;

(4) *overgeneralization*, or the abstraction of a general rule from a single event;

(5) *magnification and minimization*, by which an individual exaggerates or diminishes the importance of events; and

(6) *personalization*, an error of attribution in which one assumes either too much or too little responsibility for an event.

Cognitive distortions are not unique to depressed persons, but are apparent at all levels and types of psychiatric distress. Indeed everyone makes errors in information processing, but people who are not in distress are more able to self-correct their misperceptions.

Cognitive characteristics of suicidal individuals

The two dominant cognitive features of suicidal individuals are hopelessness and poor problem-solving ability. These characteristics appear to be independent yet interactive precursors of suicidal behaviour.

Hopelessness

This can be conceived of as a state which occurs when negative expectations are triggered. Some individuals, regardless of diagnosis, are more likely to become hopeless than others. Further, high hopelessness in one episode predicts high hopelessness in subsequent episodes (Beck *et al.* 1985). Hopelessness is more closely correlated to suicidal intent than is depression *per se* (Minkoff *et al.* 1973; Beck *et al.* 1975; Beck *et al.* 1976; Wetzel 1976; Weissman *et al.* 1979; Bedrosian and Beck 1979; Wetzel *et al.* 1980; Petrie and Chamberlain 1983; Dyer and Kreitman 1984)) and has been found to be a predictor of eventual suicide among suicide ideators (i.e., patients with suicidal ideas who have not harmed themselves) (Beck *et al.* 1985; Beck 1986).

Poor problem-solving ability

This second cognitive feature subsumes a number of specific deficits which have been found to differentiate between suicidal and non-suicidal persons, even when level of depression or severity of psychopathology is controlled. These factors are *dysfunctional attitudes, dichotomous thinking and cognitive rigidity*, and *inability to generate or act on alternative solutions*. To these research findings we would add that *viewing suicide as a desirable solution* may be considered a unique cognitive deficit.

Dysfunctional attitudes   Ellis and Ratliff (1986) found that, compared to equally depressed non-suicidal patients, suicidal patients scored higher on measures of irrational beliefs, depressogenic attitudes, and hopelessness. Bonner and Rich (1987) similarly found dysfunctional assumptions to be precursors of suicidal ideation among college students. They also found that lack of adaptive resources, such as family cohesion and reasons for living, characterized those with suicidal ideation. Thus the presence of dysfunctional attitudes may make a person vulnerable to stress, while deficient reasons for living leave the individual unable to cope.

Dichotomous and rigid thinking   Dichotomous or 'all-or-nothing' thinking categorizes events in either one of two extremes (e.g., 'I am good or bad'), and is a form of rigid thinking. It has been found to differentiate individuals with suicidal thoughts and/or suicide attempters from both normal and psychiatric control groups. On interpersonal tasks, rigid thinking prevents suicidal persons from considering alternative solutions to problems (Levenson 1974; Cohen-Sandler and Berman 1982) and from producing new ideas (Patsiokas *et al.* 1979). Moreover, suicidal patients tend to persist with their solutions even after a better strategy is presented (Levenson and Neuringer 1971).

Other problem-solving deficits   In addition to rigid and dichotomous thinking, a number of problem-solving deficits are apparent among suicidal persons. One is their general orientation to problem solving, for they tend not to accept problems as a normal part of life. Linehan and her colleagues found that suicide attempters tended to use passive problem-solving, letting others solve the problem or choosing to do nothing. Suicide ideators and non-suicidal patients were more active than suicide attempters in solving problems (Linehan *et al.* 1986). Schotte and Clum (1987) found that suicidal patients focused on anticipated negative consequences and thus failed to implement viable solutions. McLeavey *et al.* (1987) found that, compared to normal controls, suicide attempters were less able to conceptualize the means to solve a problem, generate solutions, and anticipate consequences of various courses of action. These differences were even more striking on measures with greater interpersonal content. Indeed, suicide attempters report more difficulty with interpersonal problems than do suicide ideators or non-suicidal persons (Linehan *et al.* 1986).

View of suicide as a desirable solution   In solving interpersonal problems, suicidal persons tend to become paralysed when their standard problem-solving strategy fails to work. Lacking the skills to shift to another strategy, they abandon problem-solving altogether and view suicide as the only way out. Suicide is thus an 'opiate' or relief from problems (Beck *et al.* 1979*b*).

It appears that suicidal individuals may have difficulty tolerating the anxiety of interpersonal problem solving. Suicidal children, for example, have been found to be less able than non-suicidal ones to generate comforting self-statements when faced with stressful life events (Asarnow *et al.* 1987). In addition, Linehan *et al.* (1987) report that the level of expectancy that suicide can effectively solve one's problems predicts higher suicidal intent. Thus suicide may appear to be a desirable solution when a person has faulty assumptions about its effectiveness to solve problems or when the individual is unable to tolerate the anxiety of negotiating interpersonal dilemmas.

Assessing suicidal risk

In addition to identifying cognitive processes and properties in psychopathology, research in cognitive therapy has developed assessment scales for determining suicidal risk. The Scale for Suicide Ideation (SSI, Beck *et al.* 1979a) is used to identify the salient characteristics, including the severity of the patient's intent to die, at the time of an actual attempt; and the Hopelessness Scale (Beck *et al.* 1974b) assesses pessimism or hopelessness, which is an important precursor to suicide.

In assessing suicide risk, several factors are of immediate importance. They are: level of intent to die; lethality of chosen means; access to and knowledge of how to use those means; presence or absence of some protective individual; the purpose of the attempt; and the degree of control the patient has over the desire to die; and the presence of deterrents.

Much has been written about the demographic profile of the patient likely to attempt and/or complete suicide such as age, sex, marital status, and the like. However, in clinical practice assessment of suicide risk must be based on the individual patient's motivations, goals, and deterrents. Ultimately, it is the delicate balance between the wish to live and the wish to die that determines outcome.

Acute suicide emergencies

If, in the assessment of suicide risk, it is apparent that the patient has little control over suicidal wishes, lacks support from a significant other or family member, and needs a protective environment, hospitalization is recommended. If the suicidal patient is able to function without hospitalization, it is still important to have the patient identify, and the therapist to speak with, a significant other or confidant(e) of the patient who will help monitor the patient, distract him or her from suicidal preoccupation, and alert the therapist if needed. In addition, the therapist may need to see the patient more frequently, even daily, during extremely difficult periods. If the patient is on medication, the therapist or other trusted person may keep the medication in order to prevent the patient from using it to overdose.

In emergency sessions it is important to deal directly with the patient's suicidal ideation, to undermine the patient's hopelessness, and to create disequilibrium in the patient's rigid conclusions that things cannot improve. These are the first goals of therapy (Beck *et al.* 1979*b*). By exposing faulty logic and suggesting alternative interpretations, the therapist introduces alternative explanations and solutions. At the end of the session, it is important to suggest what is to follow in the next session, helping the patient to remain open to further examination of his or her ideas, and giving the patient a reason to return to the next session.

Suicide ideation is monitored throughout treatment, for the frequency, duration, and intensity of suicidal thoughts may increase as painful topics are confronted or upsetting life events occur. Knowing the purpose of the proposed attempt directs therapy to focus, for example, on reducing hopelessness or, in the case of wishes to control others, on desires for love and affection or on maladaptive interpersonal techniques. Knowing the patient's deterrents to suicide is crucial and familiarity with the Reasons for Living Inventory (RLI, Linehan *et al.* 1983) may be helpful in adding to the patient's adaptive resources. The RLI includes survival and coping beliefs (e.g., 'I believe I can find other solutions to my problems') as well as beliefs concerning responsibility to family, child-related concerns, fear of suicide, fear of social disapproval, and moral objections to suicide.

Cognitive therapy for suicidal behaviour

Cognitive therapy is an active, directive form of psychotherapy in which the patient and therapist collaborate in the examination and modification of the patient's dysfunctional beliefs and attitudes. The patient's beliefs are treated as hypotheses to be tested, as in scientific inquiry. These 'hypotheses' can be tested logically, by exposing errors in reasoning or lack of supportive evidence, and behaviourally, by conducting behavioural tasks to challenge certain assumptions.

In treating suicidal patients, the therapist is even more active and directive, for the patient has a diminished ability to reason, remember, and create. The first goal of therapy is to undermine the patient's hopelessness by challenging cognitive distortions which make the situation appear hopeless, and by orienting the patient to a problem-solving perspective. Because of the patient's cognitive deficits, the therapist plays a dominant role initially, by generating interpretations of events, reminding the patient of contradicting evidence, and suggesting possible solutions. As therapy progresses, the patient assumes more responsibility for challenging cognitive errors, practising problem-solving, and modifying dysfunctional beliefs.

The methods of cognitive therapy for the treatment of depression are described elsewhere (Beck *et al.* 1979*b*). The following discussion reviews techniques used for cognitive deficits apparent in suicidal patients. These techniques appear in Table 3.

Table 3 *Techniques for modifying suicidal ideation*

Cognitive 'target'	Intervention
Hopelessness	Problem-solving training
	View hopelessness as a symptom
	Reduce cognitive distortions to define problems clearly
	Instigate optimism about finding solutions
View of suicide as desirable	Identify reasons for dying and reasons for living
	List advantages and disadvantages of suicide
	Bolster reasons for living and correct cognitive distortions about the advantages of dying
Dichotomous thinking	Construct a continuum between extreme points of view
	Use conditional, less absolute language
	Find 'shades of grey' in judgments
Problem-solving deficits	Minimize cognitive distortions
	Problem-solving training
Cognitive rigidity	Examine evidence in support of alternative interpretations
	Role playing with role reversal to generate alternatives

Hopelessness

Any number of cognitive distortions may complicate the patient's problems and make the situation appear hopeless. The job of the therapist is to convey to the patient that: (1) hopelessness is a state that results from negative expectations; it is not an accurate reflection of reality; (2) other interpretations and courses of action are possible; (3) the therapist believes they can work together to determine these other alternatives. The therapist and patient may list all the goals and problems about which the patient feels hopeless. At this stage, providing information or education may clarify the situation for the patient. Moreover, in order to delineate clearly the problems facing the patient, cognitive distortions must be reduced.

Cognitive rigidity

After gaining a thorough understanding of the patient's perspective, the therapist begins to create cognitive dissonance in the patient's maladaptive beliefs. This is accomplished by introducing evidence that contradicts the patient's conclusions about the self or the situation. For example, a former medical student was suicidal because she had 'made mistakes and should be punished'. Her 'mistakes' included 'letting a patient die' when she was a medical student and 'making her mother ill' when she left medical school. Examination of the evidence allowed her to see, in the first instance, that others were involved in the management of the patient in question, medical decisions are complicated and not guarantees of successful outcomes, and there is a difference between purposeful behaviour and unintended outcomes. In reference to her mother's illness, coronary heart disease, the therapist asked questions about the aetiology of heart disease so that the young woman could conclude that it is a chronic disease which is caused by many factors, not an acute reaction to stress.

Cognitive flexibility can be increased through the use of *role play, imagery, and cognitive rehearsal*. In a role play, the therapist and patient reverse roles with the patient 'assisting' the therapist in generating solutions or providing other perspectives. Another use of role reversal is to have the patient, playing his or her own negative thoughts, attack the therapist, playing the patient. The therapist logically challenges and defends against the negative statements. The roles are reversed again, with the patient practising responses to negative comments.

Visual imagery and cognitive rehearsal are used to assist a patient in anticipating problematic situations and realistically coping with them. For example, a young woman feared running into her former boyfriend and his new girlfriend at a party. She imagined that he would see she was depressed and feel superior to her. The therapist and the patient discussed possible responses to any questions about her well-being so that the patient would not feel defenceless. For example, if the boyfriend asked if she was dating anyone, she would respond briefly and directly, 'Not presently' or 'I'm keeping my options open' without discussing her loneliness. She also challenged her own negative thoughts and reminded herself, 'I was not inferior before I met him, so how can I be inferior now?' In addition, the therapist had her imagine other scenarios for the party, such as focusing on other people and initiating conversations.

Dichotomous thinking

Dichotomous thinking may result in rapid shifts in mood, making suicidal thoughts precipitous. Examples of dichotomous thinking are, 'If I can't get what I want, I'd be better off dead' and 'If she doesn't love me, I have nothing to live for.' Re-defining the situation in less extreme terms (e.g., 'I may not get what I want this time, but I can't predict the future. If I die, I'll never get what I want.') allows the patient to see additional interpretations between two extreme points of view.

Building a continuum between extreme categories allows relative judgements to be made. For example, on any given task, no one is a total success or a complete failure. Analogy and metaphor are often helpful in approaching the idea of failure. Great athletes struggle for years of training before achieving what appears to be an effortless victory. At any point before the great victory, were they failures?

Ineffective problem-solving

Problem-solving training entails guiding the patient through a sequence of cognitive and behavioural tasks: orienting oneself to problem-solving, defining the problem, generating alternatives, decision-making, implementing a solution and verifying or assessing the outcome (D'Zurilla and Nezu 1982; Hawton and Kirk 1989). It is particularly difficult for suicidal individuals to accept that problems are a normal part of life. They also have difficulty generating alternatives and seeing the long-term consequences of various courses of action. Thus the therapist needs to play a very active role, modelling problem-solving as well as collaborating with the patient in generating solutions without prejudging them.

Other cognitive distortions

Cognitive distortions interfere with clear definition of the problems facing the patient, an accurate assessment of the patient's deficits and resources, and arriving at a plan of action that the patient will actually try.

Selective abstraction is appraised by teaching the patient how to gather and observe all available information before jumping to a conclusion. Basic self-monitoring skills can be used as a way to demonstrate how the patient has ignored relevant data in the past. Identifying other sources of information is also helpful. A graduate student desperately feared that he would not get a job upon completion of his doctorate. He believed he would become destitute and concluded that it was not even worth living if that were to be his fate. The therapist asked him about fellowships and independent grants in his field. The patient had never inquired about them and was instructed to gather information before the next session.

Arbitrary inference is investigated by asking the patient to present evidence in support of his or her conclusions and then in support of more realistic interpretations. One mother who had left her child in the custody of her ex-husband at the time of divorce became suicidal years later. She concluded that she was a horrible person for having 'abandoned' her daughter in order to pursue her career. In recounting the course of events, she revealed that she had, indeed, wanted to be with the child, but that her wealthy ex-husband threatened to leave the country with the child if she tried to gain custody. The patient had been coerced into giving up. This recollection gave her more compassion for herself.

Overgeneralization can be tested by finding exceptions to the patient's rules. A patient who thought 'No one loves me, I am completely alone' was able to identify people who had expressed and demonstrated caring to him in the previous six months. He was able to see that much of his isolation was self-imposed, for he had rejected invitations to be with others.

Magnification and *minimization* are challenged by trying to establish a relative standard of comparison (e.g., How bad is this situation compared to other life events?), defining criteria on which to base judgements (e.g., 'When you say you are completely unable to cope, what do you think you need to do to cope?'), and helping patients to follow their feared scenarios through to their logical consequences (e.g., 'What if you had to leave school and tell your parents you weren't going back?'). Using this 'what if' technique allows the patient to examine the likelihood of exaggerated scenarios and consider factors which make other outcomes possible. The use of time projection (e.g., What would things be like in six months? In two years?) also helps the patient gain perspective on the relative value of an event.

Personalization is tested by having the patient identify factors, other than himself or herself, operating in a situation. After not hearing from an aquaintance for a week, a woman decided she had done something terrible and had lost the friendship. This meant, to her, that there was something wrong with her perceptible only to others. In examining other factors, the patient recognized that her friend had a busy work life, may have made plans unknown to the patient, and was typically a shy person herself. The situation was reframed as being ambiguous as opposed to a sign of rejection. The patient was then able to initiate contact with her friend.

Suicide as a 'desirable solution'

The belief that suicide is a desirable solution often rests on the assumption that it will achieve a goal without any repercussions. Thus the patient focuses on the presumed advantages of suicide, while ignoring the disadvantages of a suicide attempt. Cognitive therapists prod the patient to enumerate and examine both the advantages and disadvantages of a suicide attempt. The advantages are typically full of distortions, including misjudging the reactions of others, underestimating the amount of pain involved, and predicting the future, including the after-life. The therapist questions the advantages of a suicide attempt from scientific, religious and philosophical viewpoints, as they pertain to individual patient's belief systems.

An example of the advantages and disadvantages of a suicide attempt listed by one patient appear in Table 4. By articulating these 'pros and cons' of a suicide attempt, the patient was able to gain some emotional distance from the idea of suicide and recognize that it would not achieve her goals. Next the patient and therapist were able to contrast her reasons for living and her reasons for dying, again demonstrating that reasons for dying could be challenged logically. The therapist bolsters the patient's reasons for living by emphasising evidence that the patient may be forgetting, ignoring, or minimizing. The patient is also directed to consider the short- and long-term consequences of suicide for self and others as well as other means of achieving the goals that suicide is assumed to attain. At no time does the therapist accept suicide as an acceptable alternative.

Table 4 *Advantages and disadvantages of a suicide attempt*

Advantages	Disadvantages
I would stop being a burden to my family	My family would be ashamed and very sad
	My family would never see me get well
My pain would be over	I might maim myself and become a vegetable
Death would solve my problems	My family would have to solve the problems I leave behind
I can't go on without someone to love	If I die, no one will love me

Clinical guidelines

1. In assessing suicidal risk, intent, lethality, access to means, and the presence or absence of a protective person must all be considered.

2. Hopelessness is more closely correlated to suicide intent than is depression *per se*. The first step in treatment is to reduce hopelessness.

3. Certain cognitive deficits persist between suicidal episodes, making the patient vulnerable to further crises. These should be targets for treatment. They are rigid and dichotomous thinking, poor problem-solving skills, dysfunctional attitudes, and the view of suicide as desirable.

4. Cognitive therapy for suicidal behaviour is more active and directive than standard cognitive therapy.

References

Asarnow, J. R., Carlson, G. A., and Guthrie, D. (1987). Coping strategies, self-perceptions, hopelessness, and perceived family environments in depressed and suicidal children. *Journal of Consulting and Clinical Psychology*, **55**, 361–6.

Beck, A. T. (1986). Hopelessness as a predictor of eventual suicide. In *Psychobiology of suicidal behavior, Vol. 487*, (ed. J. J. Mann and M. Stanley), pp. 90–6. Annals of the New York Academy of Sciences, New York.

Beck, A. T., Schuyler, D., and Herman, I. (1974*a*). Development of suicide intent scales. In *The prediction of suicide* (ed. A. T. Beck, H. C. P. Resnick, and D. Lettieri), pp. 45–56. Charles Press, Bowie, MD.

Beck, A. T., Weissman, A., Lester, D., and Trexler, L. (1974*b*). The measurement of pessimism: The Hopelessness Scale. *Journal of Consulting and Clinical Psychology*, **42**, 861–5.

Beck, A. T., Kovacs, M., and Weissman, A. (1975). Hopelessness and suicidal behavior: an overview. *Journal of the American Medical Association*, **234**, 1146–9.

Beck, A. T., Weissman, A., and Kovacs, M. (1976). Alcoholism, hopelessness and suicidal behavior. *Journal of Studies of Alcoholism*, **37**, 66–77.

Beck, A. T., Kovacs, M., and Weissman, A. (1979*a*). Assessment of suicidal identation the Doalo for Guicide Ideation. *Journal of Consulting and Clinical Psychology*, **47**, 343–52.

Beck, A. T., Rush, A. J., Shaw, B., and Emery, G. (1979*b*). *Cognitive therapy of depression*. Guilford Press, New York.

Beck, A. T., Steer, R. A., Kovacs, M., and Garrison, B. (1985). Hopelessness and eventual suicide: a ten-year prospective study of patients hospitalized with suicidal ideation. *American Journal of Psychiatry*, **142**, 559–63.

Bedrosian, R. and Beck, A. T. (1979). Cognitive aspects of suicidal behavior. *Suicide and Life Threatening Behavior*, **9**, 87–96.

Bonner, R. L. and Rich, A. R. (1987). Toward a predictive model of suicidal ideation and behavior: some preliminary data on college students. *Suicide and Life Threatening Behavior*, **17**, 50–63.

Cohen–Sandler, R. and Berman, A. L. (1982). Training suicidal children to problem-solve in nonsuicidal ways. Paper presented at the annual meeting of the American Association of Suicidality, New York, April 1982.

Dyer, J. A. T. and Kreitman, N. (1984). Hopelessness, depression and suicidal intent in parasuicide. *British Journal of Psychiatry*, **144**, 127–33.

D'Zurilla, T. and Nezu, A. (1982). Social problem-solving. In *Advances in cognitive-behavioral research and therapy, Vol. 1* (ed. P. C. Kendall), pp. 201–74. Academic Press, New York.

Ellis, T. E. and Ratliff, K. G. (1986). Cognitive characteristics of suicidal and nonsuicidal psychiatric patients. *Cognitive Therapy and Research*, **10**, 625–34.

Hawton, K. and Kirk, J. (1989). Problem-solving. In *Cognitive behaviour therapy for psychiatric problems: a practical guide* (ed. K. Hawton, P. M. Salkovskis, J. Kirk, and D. M. Clark), pp. 406–26. Oxford University Press, Oxford.

Levenson, M. (1974). Cognitive characteristics of suicide risk. In *Psychological assessment of suicide risk* (ed. C. Neuringer), pp. 150–63. Charles C. Thomas, Springfield, IL.

Levenson, M. and Neuringer, C. (1971). Problem-solving behavior in suicidal adolescents. *Journal of Consulting and Clinical Psychology*, **37**, 433–6.

Linehan, M. M., Chiles, J., Egan, K. J., Devine, R. H., and Laffaw, J. A. (1986). Presenting problems of parasuicides versus suicide ideators and nonsuicidal psychiatric patients. *Journal of Consulting and Clinical Psychology*, **54**, 880–1.

Linehan, M. M., Goodstein, J., Nielson, S. L., and Chiles, J. A. (1983). Reasons for staying alive when you are thinking of killing yourself: the Reasons for Living Inventory. *Journal of Consulting and Clinical Psychology*, **51**, 276–86.

Linehan, M. M., Camper, P., Chiles, J., Strosahl, K., and Shearin, E. (1987). Interpersonal problem solving and parasuicide. *Cognitive Therapy and Research*, **11**, 1–12.

McLeavey, B. C., Daly, R. J., Murray, C. M., O'Riordan, J., and Taylor, M. (1987). Interpersonal problem-solving deficits in self-poisoning patients. *Suicide and Life Threatening Behavior*, **17**, 33–49.

Minkoff, K., Bergman, E., Beck, A. T., and Beck, R. (1973). Hopelessness, depression, and attempted suicide. *American Journal of Psychiatry*, **130**, 455–9.

Patsiokas, A. T., Clum, G. A., and Luscomb, R. L. (1979). Cognitive characteristics of suicide attempters. *Journal of Clinical Psychology*, **47**, 478–84.

Petrie, K., and Chamberlain, K. (1983). Hopelessness and social desirability as moderator variables in predicting suicidal behavior. *Journal of Consulting and Clinical Psychology*, **51,** 485–7.

Schotte, D. E. and Clum, G. A. (1987). Problem-solving skills in suicidal psychiatric patients. *Journal of Consulting and Clinical Psychology*, **55,** 49–54.

Weissman, A., Beck, A. T. and Kovacs, M. (1979). Drug abuse, hopelessness, and suicidal behavior. *International Journal of the Addictions*, **14,** 451–64.

Wetzel, R. D. (1976). Hopelessness, depression and suicide intent. *Archives of General Psychiatry*, **33,** 1069–73.

Wetzel, R. D., Margulies, T., Davis, R., and Karam, E. (1980). Hopelessness, depression and suicide intent. *Journal of Clinical Psychology*, **41,** 159–60.

7

Suicide in psychiatric hospitals: to what extent is it preventable?

GETHIN MORGAN

Psychiatric clinical services have grown and developed considerably in the last fifty years, away from a Victorian legacy of large, rigid, isolated, institutional mental hospitals in which most wards were locked, to a more flexible approach involving 'open door' small units which are more integrated with general medicine and closer to the community. No one would want to criticize these changes which undoubtedly have helped, together with more enlightened methods of care, to humanize and destigmatize the treatment of mental illness.

In spite of such changes for the better, certain unexpected findings concerning the incidence of suicide amongst psychiatric in-patients have taken us by surprise. It has of course long been known that the incidence of suicide amongst psychiatric in-patients, variously estimated to be in the order of 1.3–2.5 per 1000 discharges, is far greater than in the general population (Gale *et al.* 1980; Fernando and Storm 1984; Salmons 1984). Statistics from several countries have however demonstrated that far from falling in recent decades, the numbers of suicides occurring amongst psychiatric in-patients have actually increased progressively. The dilemma is clear: why should increased sophistication in the treatment process be associated with an increase in suicides amongst psychiatric in-patients? There is an urgent need to analyse why this has occurred and to identify deficiencies in our present approach that may be causal and which need to be rectified.

Recent evidence and clinical experience

The nature of the evidence

Even in psychiatric wards, suicide is a relatively rare event. The study of such suicides by individual clinicians and even single hospitals is inevitably based retrospectively on low numbers of sporadic cases, from which it is rash to generalize. It would be extremely difficult to mount controlled prospective surveys of suicide amongst psychiatric in-patients, yet without them it is hazardous to comment on cause and prevention; we can never be certain in routine clinical practice whether we ever prevent a suicide or even precipitate one. National statistics are of course valuable in providing an overview, but their deficiency should always be kept in mind. Thus if a patient should die in another hospital or soon after discharge, then the event is not related in such statistics to the hospital from which recent discharge or transfer had occurred. Such are the deficiencies in the basic data upon which studies of suicides in hospital have inevitably to be based, and so much greater must be the hazards of drawing conclusions from them.

Statistics of hospital suicides

Crammer (1984) has drawn together the findings from studies in several countries and alerted psychiatrists to the worrying fact that during the period 1950–74 the in-patient suicide rates in Finland, Norway, Sweden, and Switzerland increased significantly, even though the rates for the general population remained unchanged.

More recently Sainsbury (WHO 1982, p. 25) has commented on a similar picture both in the Netherlands and England. In the latter, suicide rates fell in the period 1920–56, with the advent of open hospitals and liberal treatment policies, but subsequently have increased progressively from 37.5 to 84.0 per 100 000 during the period up to 1973 (Table 5).

Such statistics have led to much speculation concerning the possible causal role of factors such as the 'open door' approach, changes in admission policy, the abandonment of a hierarchical authoritarian system of hospital care in favour of a diffusion of responsibility by multidisciplinary consensus, and permissive management. Before proceeding further, however, it should be noted that there are problems in defining the base population. Undoubtedly the development of psychiatric services in recent years has been characterized by a shorter in-patient stay with more rapid patient turnover. The mean average number of patients in psychiatric hospitals is in itself inappropiate as a denominator because it does not allow for this vastly increased rate of exposure of individuals in the hospital

Table 5   *Suicide rates: psychiatric in-patients (England)*

Period	Mean number of patients	Mean number of suicides	Annual suicide rate (per 100000)
1920–21	101 438	48.7	48.0
1945–47	133 428	68.7	51.5
1954–56	146 847	55.0	37.5
1964–66	128 728	76.7	59.6
1972–73	111 601	93.8	84.0

Source: WHO (1982)

environment. The rates may therefore be artificially inflated due to an erroneously low denominator. We should also keep in mind the possibility that persons admitted to psychiatric wards in recent years may themselves be more suicide prone, and remain so for a greater proportion of time during their stay than was the case previously. A recent analysis of hospital statistics concerning psychiatric patients in various European countries claims that no increase in suicide rate has occurred; the rise in numbers of suicides in this group is entirely related to a much greater patient turnover and increased risk propensity of the patients themselves (Wolfersdorf *et al.* 1988).

The utmost caution should be observed therefore before assuming that changed patterns of care may have caused an increased rate of suicides amongst psychiatric in-patients. Nevertheless one fact is undeniable: the event of suicide has become more common in our psychiatric wards. That in itself needs to be addressed.

Which psychiatric patients?

If suicides occur under our very noses, as Sainsbury (1988) has put it, why do we fail to detect those psychiatric in-patients who will kill themselves? How do they elude us?

On the face of it, the high risk groups can be identified fairly clearly amongst psychiatric in-patients (Hawton 1987). Whilst no diagnostic entity can be safely excluded, the front runner in terms of suicide risk is depressive illness, accounting for about a half, the remainder consisting of personality disorders, alcoholism and drug abuse, schizophrenia, and organic brain syndromes. Furthermore, there is now a considerable body of data concerning clinical clues that distinguish individuals in these diagnostic groups who proceed to suicide from others who do not (Table 6).

Data from a study of suicides in the community suggest that depressives who commit suicide, when compared with a control group (Barraclough and Pallis 1975), have more persistent insomnia, manifest more self-neglect, have more impairment of memory, are more agitated, and most characteristic of all, have made significantly more previous suicide attempts; they are also older, more often male, single or separated, and more socially isolated (42 per cent lived alone compared with 7 per cent of controls). These findings must surely have relevance to psychiatric in-patients, although some difference may occur; thus Copas and Robin (1982) have shown that the effect of increasing age may not be significant amongst the female in-patient group. Feelings of hopelessness are however undoubtedly important indicators of increased suicide risk, whatever the setting (Beck *et al.* 1975; 1985).

Most schizophrenics who kill themselves tend to be young, male, with a history of previous attempts at suicide, and episodes of depression with anorexia and weight loss. High intellectual ability and educational

Table 6   *Diagnostic groups: suicide risk*

Diagnosis	Risk factor
Depression:	Male, older, single/separated, socially isolated
	Previous deliberate self-harm (DSH)
	Persistent insomnia, self neglect, impaired memory, agitation
Schizophrenia:	Male, younger
	Previous DSH and depressive episodes with anorexia/weight loss
	More serious illness, recurrent relapse
	Fear of deterioration, especially in those of high intellectual ability
Alcohol/drug addiction:	Adverse life events
	Previous DSH
	Depressed mood, serious physical complications

Source: Barraclough and Pallis 1975; Drake *et al.* 1984, 1985; Roy 1986

achievement with fear of psychological deterioration is another common theme (Drake *et al.* 1984, 1985). The schizophrenic illness also tends to be more severe with recurrent relapses (Roy 1986). Alcoholism and drug abuse account for up to 20 per cent of in-patient suicides, and here again the suicides tend to have made more previous attempts, to have been depressed, and to experience more adverse life events and serious physical complications (Murphy and Robins 1967; Murphy *et al.* 1979).

All these studies emphasize that clinical assessment should take account of situational and demographic as well as individual clinical factors if suicidal individuals are to be identified effectively. Beside this we can also use a series of predictive scales which evaluate suicide risk (Burk *et al.* 1985). The value of such instruments is undoubtedly greatly enhanced when their use is focused on clinically identified high-risk groups, such as those found within the psychiatric in-patient population.

How valuable then is such a rich diversity of data in predicting individual suicides in hospital wards? Sadly, several factors would make the exercise less than perfect.

First of all we must remember that the clinical stereotype of suicide is probably changing. In recent years the well recognized association of increased suicide risk with older age in deliberate self harm, for example, has become far less obvious and the incidence of suicide in young adults, especially males, is increasing. The other dilemma here has been discussed recently by Kreitman (1988) who reminds us that even in groups of patients at exceptionally high risk of suicide, its incidence is still low; for example, in those who have deliberately harmed themselves, the proportion who commit suicide within 12 months of hospital admission is no more than 1–2 per cent. In the long term the proportion becomes greater but the low base rate means that predictions at an individual level within the immediate future is extremely difficult. Prediction scales can achieve a reasonable degree of sensitivity over an extended period of time, but they are relatively ineffective in the acute clinical situation, where they are effective only at the expense of low specificity, so that they include a large number of false positives.

We are forced to conclude that high risk clinical stereotypes and predictive scales can offer only limited help in our routine clinical work, which demands decisions based on immediate suicide risk. Well identified risk factors only occur in the minority of actual suicides and as Kreitman (1988) has commented: 'It would be clinically disastrous to assume that the absence of risk factors means that the possibility of a fatal outcome can be ignored, since by doing so most cases can be missed.'

Interviewing the suicidal

A further area relevant to reasons why suicides occur in psychiatric wards concerns problems which beset face-to-face assessment of those at risk. Undoubtedly there are many pitfalls here which may mislead the unwary.

In reviewing a personal series of 12 psychiatric in-patients who had committed suicide either during their in-patient stay or very soon afterwards, Morgan (1979, p. 50) attempted to pinpoint reasons why assessment had failed to predict suicide (Table 7). These included misleading variability in the degree of distress and expressed suicidal ideation, related to an ambivalence that is probably characteristic of most suicidal individuals but also often dependent upon the degree of contact with ongoing traumatic life events. In some instances angry and provocative behaviour had led to a progressive deterioration in their relationship with others, including ward staff; a process which Morgan (1979) has termed 'malignant alienation'. Such a development, in which the patient is seen as manipulative or even malingering, means that the potential for constructive therapy rapidly disappears. Such difficult and uncooperative behaviour of suicidal patients has long been documented (Seager and Flood 1965). In a further series of psychiatric patients who committed suicide derived from coroners' records, Morgan and Priest (1984) identified a further factor, namely misleading clinical improvement in the ward situation, apparently because of removal from stressful life events which nevertheless remained unresolved, and which had catastrophic effects when the patients were discharged home. Misleading improvement in 23 of 57 suicides that occurred in various hospitals in Birmingham was thought to be related to resolution of internal conflict as a result of a clear decision to die (Goh *et al.* 1989).

Table 7   *Suicidal patients: hazards in assessment and management*

Denial of suicidal ideas	Spurious assumptions by staff
Variability in degree of distress	Surveillance problems
False improvement	Poorly planned treatment programme
Anger	
Uncooperative behaviour	
Malignant alienation	

Source: Morgan 1979; Morgan and Priest 1984

Faulty interview techniques on the part of staff may also be a significant problem. It is necessary to lead into questioning about suicidal ideas very gradually; aggressive interrogation will certainly be unreliable. False assumptions by staff are also common. For example, fear of implanting suicidal ideas may mean that any enquiry concerning them is avoided, or a patient who openly talks of suicide is then necessarily regarded as being free from risk; both of these are common and serious errors. Further problems may stem from inappropriate staff attitudes to suicide, such as regarding some patients as being so sick, and their problems so hopeless, that their suicide is inevitable; such patients are not slow to note any resulting lack of enthusiasm in their care. Similarly a staff member who entertains the idea that suicide is in certain circumstances justifiable and so connives with it might convey this belief, perhaps unwittingly, to the patient.

Ward environment

The physical hazards that present themselves in any hospital situation are obviously important. Periods of special danger are soon after admission (Copas and Robin 1982), during leave, or soon after discharge from the ward. Crammer (1984) has reviewed various studies which reveal that a third of suicides occur whilst the patient is on leave; even more of the actual suicides occur away from the ward itself. This again highlights the very common finding that such suicides often take us by surprise, presumably because we have been mistaken in our clinical assessment of current risk.

The phenomenon of clustering of suicides in hospital psychiatric in-patients has from time to time gained much publicity. Intense comment and reaction, especially in the media, tends to inflame the situation, and poor quality data render it difficult to evaluate the true significance of such clusters of suicides. Effective preventive measures must nevertheless keep in mind this 'contagion' element, whereby one or more suicides in a relatively closed community such as a hospital might lead, through processes which are ill understood, to kindling of suicidal ideation in other vulnerable individuals. Good practice should encompass the needs of the ward community as a whole, as well as those of vulnerable individuals when a suicide has occured in its midst. Clusters of non-fatal deliberate self-harm also occur and may be associated with an increased risk of actual suicide in those involved (Hawton 1978).

Patient management: medical and physical treatment

Barraclough *et al.* (1974) have analysed the irregularities in the way medication has been prescribed by general practitioners in the case of 100 patients who committed suicide in the community, and such findings inevitably have significance for hospital practice. They found that only half of depressed suicides were receiving antidepressant drugs, and quite often the level of dosage was inadequate; the inappropriate use of depot phenothiazines for patients who did not suffer from schizophrenia also seemed a significant problem. Sainsbury has suggested that the increase in suicide in patients soon after discharge may be related to premature discontinuation of medication. He also urges that ECT should be retained as a very effective way of treating severe depression and he deplores the manner in which lobbyists have in some quarters prevented patients from enjoying its benefits (Sainsbury 1988).

Relationships with staff and levels of observation

In order to be effective in suicide prevention a hospital ward should provide an appropriate milieu in which the patient has to live. A settled, calm, confident environment is the most appropriate setting for the delivery of effective therapy for any individual patient. The degree of disturbance in other patients, the level of comfort, the warmth, maturity, and professionalism of staff, provision of a key therapist, sustaining close relationships, the mutual trust between doctor and nurse—all these must be relevant to the degree of suicide risk in an individual patient. There is, however, one dilemma that has never been fully resolved, and this concerns the advisability of providing close observation of patients at risk of suicide. How effective is it and is it ever counter-productive? The old system of suicide caution has been rightly criticized as it deprived the patient of all sense of privacy, and possibly increased suicide risk. Yet our open ward system may well itself be hazardous without some graded system of observation for persons at risk of suicide.

The problems reported by Gardner (1988) are likely to apply widely. Attempts to agree such a scheme of graded levels of observation according to the degree of suicide risk can produce a great deal of tension, indeed rivalry, between professional groups as to how it should be implemented, who should make the decisions and who can change them. Some professional staff seem to prefer a rule-free situation, ostensibly to maintain a flexible approach, but possibly also related to the fear of litigation if things go wrong. Nursing officers and managers have the problem of providing sufficient numbers of staff and they too might prefer a situation in which they are not hidebound by rules. If we accept that a clear, explicit set of guidelines is required concerning the supervision of suicidal patients then all these issues must be faced in any psychiatric unit, and healthy trusting relationships between nurses and doctors are an essential ingredient in their resolution.

Even when the principle has been agreed, the details of an observation scheme for suicidal patients are not easy to finalize. Many deficiencies exist in current practice. Schemes may vary from one ward to another even within a single psychiatric service and may involve terms such as 'close', 'special' which are ambiguous and far from explicit. The principles however are clear: the scheme should be understood by all staff on the ward, decisions should be documented in both nursing card index notes as well as medical records, they should be decided upon as a result of discussion by both nurse and doctor, and they should be formally reviewed by the whole clinical team at least weekly. The scheme should mean that the patient benefits from close one-to-one contact with identified key workers rather than be subjected to impersonal surveillance.

The extent to which the rules can be interpreted flexibly should be made clear. Everyone should be aware of certain danger points, when it is tempting to relax the rules, because it is then that suicides unexpectedly occur; for example whether or not to leave a heavily sedated patient fully clothed temporarily on admission, or allow the curtains to be drawn around the bed at night—such considerations may seem trivial at the time but in reality can be potentially dangerous in the case of high-risk individuals.

Practical guidelines in the prevention of suicide of psychiatric in-patients*

Without doubt some suicides amongst psychiatric in-patients are inevitable, no matter how scrupulous the clinical care. Nevertheless others may be preventable, and an agreed code of practice is essential, if only to increase our awareness of the many inherent hazards and pitfalls in caring for such patients.

Ten basic elements of good clinical practice which may help to prevent suicide amongst psychiatric in-patients, and which arise out of the evidence which has been evaluated in this chapter, are now set out in the form of a checklist.

1. *Ward milieu*

Facilities should be adequate in terms of physical safety with reasonable control of specific hazards, such as high windows and staircases. Sufficient security means an appropriate control of ward exits and the provision of an

---

*The ten-point Code of Practice outlined below is © H. G. Morgan, 1990.

area in the ward setting which allows complete control of access and exit for any patients requiring intensive supervision. A guaranteed minimum agreed level of staffing is essential. If any ward cannot meet the needs of high risk individuals, then a ward which can provide more intensive care should be available to which they can be transferred as necessary. Ideally extra staff should be made available, thereby avoiding the disruption of moving from one ward to another. The locked ward concept is now only appropriate in a forensic setting, because in normal circumstances intensive levels of supervision should depend upon appropriate provision of staff rather than impersonal physical barriers.

Any acute ward is bound to experience disturbance from time to time. Suicide prevention will inevitably be related to a positive atmosphere in which clinical crises are dealt with as rapidly as possible in a quiet, efficient way with the least possible disturbance to other patients. Wards that are too large, for example containing more than 25 patients, make it difficult to deliver personalized care for individuals. Insufficient numbers of beds means that the atmosphere becomes unsatisfactory because of constant manoeuvring of patients, sometimes involving lodging them out to other wards with little warning.

High patient morale depends upon adequate provision of basic facilities and good communications not only between patient and therapist on an individual basis but also with regard to the ward community as a whole, whether through regular ward meetings or otherwise.

2. *Relationship with hospital managers*

The principles of good care for suicidal patients should be understood fully by the hospital managers, particularly in view of the fact that it may have significant resource implications. In-service training of managers should include basic clinical matters to ensure that they understand the principles on which patterns of care are based.

3. *Staff complements*

Nurses are, of course, in contact with in-patients for many hours of the day. The positive role they can play in suicide prevention is immense, and is closely related not only to their professional skills and experience but to the morale which should permeate their work. In turn this depends upon regular availability of appropriate numbers of qualified staff. Indeed, unless basic issues such as staffing levels are satisfactory, the delivery of care to special need groups such as the suicidal will remain rudimentary.

4. *Staff relationships*

Important too is the level of mutual trust that should exist between professional groups, especially nurse and doctor; each has to learn and maintain respect for the other. When faced with patients who exhibit difficult behaviour, a situation akin to rivalry can develop concerning who has the right to make decisions or countermand them; this does nothing in the cause of suicide prevention yet is more common than we might care to acknowledge.

5. *Staff education and attitudes*

Appropriate student nurse training, ideally by ward medical staff, concerning the basic issues of care for the suicidal, including first aid, should occur very early in their ward attachments. Ongoing in-service training for qualified nurses should also be regularly available. Junior doctors should receive instruction concerning care of the suicidal at the earliest possible time following their entry into psychiatric training. Their seniors should emphasize the principle that good care depends on the fullest possible communication between nurse and doctor on a regular basis.

6. *Admission procedures*

The doctor who agrees to admit a patient who is at increased risk of suicide should communicate immediately with the nurse in charge of the ward indicating the level of risk and appropriate supervision which is required following that patient's arrival on the ward and prior to examination by the ward doctor. Any significant suicidal risk should mean that a new patient must remain in bed in pyjamas and should be observed intensively until the admitting doctor carries out a full assessment. There should be universal awareness that the period soon after admission presents a particular hazard with regard to suicide risk.

7. *Clinical assessment procedures*

Thorough, well-documented, clinical assessment based on a clinical history and examination must remain the mainstay of good clinical practice in the care of the suicidal. For example, taking the trouble to obtain notes from another hospital where the patient may have been admitted previously, and interviewing other informants may seem obvious but so often is not done. The use of specialized risk questionnaires can be a useful back-up to thorough clinical assessment, but no more than that. Instruments such as the Beck Depression Inventory (Beck *et al.* 1961) and Hopelessness Scales (Beck *et al.* 1974) can be particularly useful for monitoring a patient's condition throughout in-patient stay.

   The admitting doctor should communicate immediately with the nurse in charge once the admission interview has been completed, and a treatment plan together with appropriate levels of observation agreed. Nurses should be encouraged to feel that they form an integral part of risk assessment and their views on level of risk are very important; their contact with patients is the closest and most extended of all professional groups and their opinions must as a result be greatly enhanced in reliability.

8. *Clinical management procedures*

Each hospital unit should have an agreed code of practice concerning graded levels of observation of suicidal patients depending upon the judged degree of suicide risk. Such a scheme should be shared by all units throughout the health district concerned. It should use terms that are explicit, descriptive, and easily understood. Each decision should be a joint one, based on discussion between the appropriate doctor and nurse; any change in level of observation should again be a matter for joint consultation, although in emergencies nursing staff can increase the level if necessary; all staff in the unit should understand the scheme, and be fully informed about that which currently applies to any patient judged to be at risk. Decisions should be reviewed daily by the ward doctor and senior nurse, and reviewed by the multidisciplinary team at least weekly. Such a scheme should aim to foster a patient's sense of close personal support rather than constitute an impersonal surveillance that could be counter-productive. It should aim in particular to prevent the sequence of alienation that can occur terminally in the case of some patients who actually proceed to commit suicide. For patients who are convalescing, decisions regarding unsupervised periods away from the ward (for example parole in the hospital grounds or leave in the community) should be taken with the utmost care.

9. *Relationship with other hospital units and the community*

Communications with health care professionals and relatives in the community should be two-way, full, regular, easy to achieve and never delayed. When a patient is about to be discharged, the general practitioner in particular should be fully informed concerning medication and further treatment plans. Any transfer to other units also should occasion full and effective exchange of communication between the staff involved with regard to the level of risk and the precautions taken to deal with it. It is important that the patient too should be well prepared for any upheaval such as ward transfer. Good relationships with other agencies such as the police, coroner, and the media depend upon understanding based on good two-way communication.

10. *Suicide audit and review procedures*

Readiness to review and so learn from adverse events is surely one measure of a hospital ward's quality of care. Regular reviews of untoward events and suicide audit procedures are an example of such good clinical practice, though they can be difficult to set up, partly because the emotional trauma in hospital makes formal review extremely upsetting for all concerned. Nevertheless they are an important feature of good clinical practice. Their aim should not be to apportion blame but to help staff to learn from experience. By providing group support they should build up sufficient confidence amongst staff members to allow an effective retrospective scrutiny of events, unclouded by defensive reconstruction.

References

Barraclough, B. M. and Pallis, D. J. (1975). Depression followed by suicide: A comparison of depressed suicides with living depressives. *Psychological Medicine*, **5**, 55–61.

Barraclough, B. M., Nelson, B., Bunch, J., and Sainsbury, P. (1974). A hundred cases of suicide: clinical aspects. *British Journal of Psychiatry*, **125**, 355–73.

Beck, A. T., Ward, C. H., and Mendelson, M. (1961). An inventory for measuring depression. *Archives of General Psychiatry*, **4**, 561–71.

Beck, A. T., Weissman, A., Lester, D., and Trexler, L. (1974). The measurement of pessimism: The Hopelessness Scale. *Journal of Consulting and Clinical Psychology*, **42**, 861–5.

Beck, A. T., Kovacs, M., and Weissman, A. (1975). Hopelessness and suicide behaviour: an overview. *Journal of the American Medical Association*, **234**, 1140–9.

Beck, A. T., Steer, R. A., Kovacs, M., and Garrison, B. (1985). Hopelessness and eventual suicide: a 10 year prospective study of patients hospitalized with suicidal ideation. *American Journal of Psychiatry*, **145**, 559–63.

Burk, F., Kurz, A., and Moller, H-J. (1985). Suicide risk scales: Do they help to predict suicidal behaviour? *European Archives of Psychiatry and Neurological Science*, **235**, 153–7.

Copas, J. B. and Robin, A. (1982). Suicide in psychiatric in-patients. *British Journal of Psychiatry*, **141**, 503–11.

Crammer, J. L. (1984). The special characteristics of suicide in hospital in-patients. *British Journal of Psychiatry*, **145**, 460–3.

Drake, R. E., Gates, C., Cotton, P. G., and Whitaker, A. (1984). Suicide among schizophrenics: who is at risk? *Journal of Nervous and Mental Disease*, **172**, 613–17.

Drake, R. E., Gates, C., Whitaker, A., and Cotton, P. G. (1985). Suicide among schizophrenics: a review. *Comprehensive Psychiatry*, **26**, 90–100.

Fernando, S. and Storm, J. (1984). Suicide among psychiatric patients of a district general hospital. *Psychological Medicine*, **14**, 661–72.

Gale, S. W., Messnikoff, A., and Fine, J. (1980). A study of suicide in state mental hospitals. *Psychiatric Quarterly*, **52**, 201–13.

Gardner, R. (1988). Surveillance of patients at risk. In *The clinical management of suicide risk*, ed. H. G. Morgan (pp. 21–3). Conference held at The Royal Society of Medicine. Chapterhouse Codex Ltd, Hants.

Goh, S. E., Salmons, P. H., and Whittington, R. M. (1989). Suicide in psychiatric hospitals. *British Journal of Psychiatry*, **154**, 247–50.

Hawton, K. (1978). Deliberate self poisoning and self injury in the psychiatric hospital. *British Journal of Medical Psychology*, **51**, 253–9.

Hawton, K. (1987). Assessment of suicide risk. *British Journal of Psychiatry*, **150**, 145–53.

Kreitman, N. R. (1988). Some general observations on suicide in psychiatric patients. In *The clinical management of suicide risk*, pp. 7–9. Conference held at The Royal Society of Medicine. Chapterhouse Codex Ltd, Hants.

Morgan, H. G. (1979). *Death wishes: the understanding and management of deliberate self-harm*, Wiley, Chichester.

Morgan, H. G. and Priest, P. (1984). Assessment of suicide risk in psychiatric inpatients. *British Journal of Psychiatry*, **145**, 467–9.

Murphy, G. E. and Robins, E. (1967). Social factors and suicide. *Journal of the American Medical Association*, **199**, 303–8.

Murphy, G. E., Armstrong, J. W., Hermele, S. L., Fischer, J. R., and Clenendin, W. W. (1979). Suicide and alcoholism: interpersonal loss confirmed as a predictor. *Archives of General Psychiatry*, **36**, 65–69.

Roy, A. (1986). Suicide in schizophrenia. In *Suicide* (ed. A. Roy) (pp. 97–112). Williams and Wilkins, Baltimore.

Sainsbury, P. (1988). Suicide prevention: an overview. In *The clinical management of suicide risk*, (ed. H. G. Morgan) (pp. 3–6). Conference held at The Royal Society of Medicine. Chapterhouse Codex Ltd, Hants.

Salmons, P. H. (1984). Suicide in high buildings. *British Journal of Psychiatry*, **145**, 469–72.

Seager, C. P. and Flood, R. A. (1965). Suicide in Bristol. *British Journal of Psychiatry*, **111**, 919–32.

Wolfersdorf, M., Keller, F., Schmidt-Michel, P. O., Weiskittel, C., Vogel, R., and Hole, G. (1988). Are hospital suicides on the increase? *Social Psychiatry and Psychiatric Epidemiology*, **23**, 207–16.

World Health Organization (1982). *Changing patterns in suicide behaviour*. Euro Report and Studies 74. WHO, Copenhagen.

8

Self-cutting: can it be prevented?

KEITH HAWTON

The problem

Broadly speaking there are three types of self-injury in which cutting occurs. These are:

(1) *superficial self-cutting*, usually of the wrist or arm, but which may involve other areas of the body, and is usually associated with little or no suicidal intent;

(2) *deep cutting*, which may endanger major blood vessels, nerves, and tendons, and is usually, but not always, associated with serious suicidal intent;

(3) *self-mutilation*, which often results in disfigurement, may or may not endanger life, and usually occurs in individuals with psychotic illness.

While there is some overlap between these three categories and more than one type of injury may be found in the same individual, it is reasonable to consider them separately, especially with regard to management. Superficial self-cutting, which is by far the most common, is the focus of this chapter.

Patients who cut themselves are encountered in several settings. Among 'attempted suicide', 'parasuicide', or 'deliberate self-harm' referrals to accident and emergency departments, some 10–15 per cent will be cases of self-injury (Weissman 1975; Hawton and Catalan 1987, p. 150) and of these the vast majority will be patients who have cut themselves, mostly superficially. Self-cutters are also at increased risk of taking overdoses, although the motivation for the two types of act may differ. Repetition of self-cutting occurs frequently. Self-cutting is especially common in psychiatric units (Hawton 1978), where it may occur in an apparently epidemic or contagious fashion (Simpson 1975), and in prisons. It is impossible to estimate the full incidence of self-cutting because it occurs in diverse settings and many episodes do not come to medical attention.

As Simpson (1976) noted, 'This relatively common condition probably constitutes one of the most difficult treatment and management problems in clinical psychiatry'. Unfortunately, there are few guidelines to assist the clinician. No properly controlled treatment studies of self-cutting have been reported. In order to understand why prevention of self-cutting is so challenging and to obtain clues about management strategies which might be helpful, it is important to be aware of the characteristics of people who cut themselves and the motivational nature of self-cutting.

Characteristics of people who cut themselves

Typical self-cutters have been described as young, single, often attractive, females. While it is true that self-cutters are mostly young, the focus on females is probably misleading, and may reflect the impressions of psychiatrists working in specialized settings. In fact, roughly equal numbers of male and female self-cutters are seen in the accident and emergency departments of general hospitals (Weissman 1975; Hawton and Catalan 1987, p. 151). However, it appears that many more female self-cutters enter treatment; therefore the female pronoun has been used in this chapter.

Backgrounds   The following factors are relatively common in the histories of self-cutters: broken homes, although probably no more frequently than in overdose patients (Rosenthal *et al.* 1972); lack of parental emotional warmth and physical contact, especially from mothers (Simpson 1975); hospitalization and surgery before the age of five years (Rosenthal *et al.* 1972; Simpson 1975); and childhood physical and sexual abuse (Favazza 1987). An excessive proportion of self-cutters have worked in paramedical fields, especially nursing (Grunebaum and Klerman 1967; Simpson 1975).

Personality   Most self-cutters have persistently low self-esteem, and many actively dislike their bodies. They may have considerable difficulty in expressing their emotional needs or experiences (Simpson 1976). Some authors have suggested that self-cutters often have obsessional personality traits (Gardner and Gardner 1975).

Psychiatric disorders   It is essential to regonize that self-cutting is not itself a psychiatric diagnosis, although some workers have attempted to delineate a 'deliberate self-harm syndrome' (Pattison and Kahan 1983). This has six features:
(1) sudden and recurrent irresistible impulses to harm oneself without the perceived ability to resist;
(2) a sense of existing in an intolerable situation which one can neither cope with nor control;
(3) increasing anxiety, agitation, and anger;
(4) constriction of cognitive processes resulting in a narrowed perspective on one's situation and personal alternatives for action;
(5) a sense of psychic relief after the act of self-harm; and
(6) a depressive mood, although suicidal ideation is not typically present.

While these criteria highlight some of the typical characteristics of self-cutting, distinguishing this phenomenon as a syndrome may not be especially helpful. It may be more appropriate to regard self-cutting as a symptom, just as one might regard binge-eating or taking an overdose.

Many self-cutters are prone to rapid mood swings, set-backs tending to precipitate a sudden sense of despair and hopelessness. Cutting may also occur in association with actual episodes of clinical depression in which there are the characteristic psychological and physical affective symptoms.

In the USA, self-cutters will often receive the DSM-III diagnostic label of 'borderline personality disorder', recurrent self-mutilation behaviour being one of the criteria for this diagnosis (American Psychiatric Association 1987).

Other associated problems   Many self-cutters abuse alcohol or drugs, and their consumption may increase the likelihood of a cutting episode. There is also a significant association with eating disorders, including both binge-eating and anorexia, these being found in 65 per cent of one series (Rosenthal *et al.* 1972) and in 75 per cent of another (Simpson 1975). In many ways self-cutting and binge-eating share similar characteristics, and may occur at the same time in some subjects. Menstrual disturbances, including negative reactions to the onset of menstruation, dislike of menstruation, and menstrual irregularities (Rosenthal *et al.* 1972; Simpson 1975), and sexual problems, including difficulty in sustaining sexual relationships, gender role confusion, and sexual dysfunction (Simpson 1976), are also common in self-cutters.

The nature of self-cutting

Pattern of cutting   Cuts may be single or multiple, some cutters making multiple superficial incisions. The forearms are the most common site, but the cuts may be made to any part of the body which is accessible. Cutting is frequently repeated, some individuals cutting themselves many times. Most acts draw blood, razor blades and broken glass being the favoured cutting instruments. Other types of self-mutilation may occur, burning with cigarettes being quite common.

Cutting often follows an event which activates a sense of loss or abandonment. The event may sometimes appear to the outsider to be relatively trivial. However, in order to understand the impact of the event on the individual it is necessary to gain insight into the personal meaning of the event, and especially how it relates to the individual's attitudes towards herself and how she perceives her role in personal relationships.

The act of cutting   The individual may describe mounting feelings of tension and anger, and often a sense of hopelessness. The urge to cut, once activated, appears to become irresistible. Cutting is usually carried out alone. Immediately before cutting the individual may feel numb or empty, and often depersonalization or dissociation occur.

The majority of cutters experience no pain while cutting themselves, and virtually all describe a profound sense of relief immediately afterwards. The sight of blood may be important in this process. However, the sense of relief is often mingled with feelings of disgust and guilt.

Suicidal intent is usually absent in this type of self-cutting, the most common motives being relief from tension and expression of anger towards the self or others (Gardner and Gardner 1975).

There has been remarkably little research on the factors which maintain the behaviour. Four models might be relevant, each of which has somewhat different treatment implications:

(1) a social reinforcement model, in which the cutting is a means of communication (expressing anger) to other people;

(2) an analogue of obsessive-compulsive behaviour;

(3) a means of ending an abnormal aversive psychological state; and

(4) a means of self-punishment.

The available evidence suggests that all these models may be of relevance, but to varying degrees in different individuals.

Effects on other people   Most people experience a sense of revulsion and incomprehension when confronted by self-cutting. The behaviour presents special problems for staff on in-patient psychiatric units who may feel impotent to prevent it, threatened by their lack of sense of control over the situation, and angry at what might be interpreted as a rejection of their therapeutic efforts. Furthermore, conflict may arise between staff members concerning the restrictions which may be placed on the patient, the therapeutic programme which an individual staff member is following with the patient, and often the basic question of whether the patient should be in hospital at all. Unless these conflicts can be resolved they may further contribute to the patient's difficulties.

Management

In trying to help self-cutters, it is important to remember some of the key features which have been discussed, especially the rapidity of mood swings, the difficulties many of these patients have in verbalizing their feelings, the extent of the tension which may precede self-cutting, the strongly reinforcing nature of self-cutting, and the basic sense of low self-worth and personal dislike which many cutters experience. Five aspects of management will be considered, namely assessment, helping the patient to gain control over cutting, helping with underlying difficulties, the use of medication, and management strategies in the in-patient setting.

Assessment

Before any attempt is made to help a patient gain control over self-cutting, it must be established that the patient wishes to try to give up the behaviour. However, in view of the powerful emotions that precede an act of cutting and the considerable relief it brings, it is not unusual for a patient initially to appear ambivalent about stopping cutting because she can see no other way to control her emotions.

The assessment should include a careful and detailed *behavioural analysis* of the sequence which leads to self-cutting. This should focus on the factors shown in Fig. 3. Such an analysis should help both the therapist and the patient develop a fuller (if incomplete) understanding of the behaviour and may help identify strategies which could either break the chain which leads to self-cutting or replace cutting itself.

Identification of the cognitions which lie behind the development of tension and other emotions which precede cutting sometimes requires considerable effort, perhaps through encouraging the patient to imagine in detail a typical provoking event and the consequences (behavioural, emotion, and cognitive) which might lead to cutting. The patient may initially only be aware of the emotions experienced since the thoughts may be 'automatic' (overlearned, fleeting, and hardly entering conscious awareness). Explaining how emotional states in response to events are the result of such thoughts (Beck *et al.* 1979) may help in this process. These thoughts are likely to relate to more general attitudes or assumptions the patient holds concerning herself; however, careful exploration during subsequent treatment sessions may be necessary to identify these.

Helping the patient gain control over cutting

Because the prevention of self-cutting has received so little empirical attention, the guidelines on management which follow represent an assembly of strategies from which clinicians may choose when faced with individual cases.

The patient should first be asked to look at the events which typically provoke the emotional state which leads to cutting to see if these can be avoided. Usually this will not be possible since the events are likely to be those which are readily encountered in everyday life. It may be more profitable to look at aggravating factors (e.g., alcohol consumption) in terms of the extent to which these can be controlled, and factors which decrease the likelihood of cutting (e.g., contact with other people) and the extent to which these can be utilized.

Next, the therapist might help the patient review whether there are any ways in which the rise in tension which is so often associated with cutting can be reversed. With regard to the underlying cognitive processes, it is useful to identify the immediate contributory thoughts but to explain to the

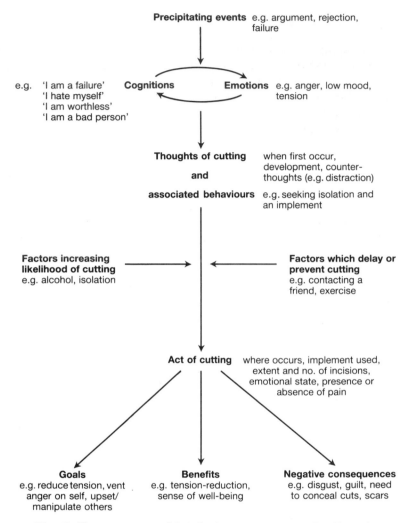

Fig. 3. Factors to consider during assessment of self cutting.

patient that these can be worked on as therapy proceeds, and then to look at specific strategies to help the patient control her cutting. It is important to try to find ways by which cutting can be controlled, rather than focusing all therapeutic attention on underlying factors, because cutting generally adds to the patient's sense of personal disgust and low self-worth, thus increasing the likelihood of further episodes of self-cutting. Furthermore, if the patient can begin to gain a sense of control over the cutting this can be an important step in therapy because it may foster self-confidence and optimism.

Relaxation In view of the build-up of tension which so often precedes self-cutting, progressive muscular relaxation or the newer but similar technique of applied relaxation (Ost 1987) would appear to be obvious methods to try. However, most patients describe the rise in tension before cutting as so powerful that relaxation simply does not work. Nevertheless, it is worth teaching the patient relaxation to use in the early stages of increasing tension, especially later in therapy when other therapeutic strategies may have brought about some amelioration in the problem.

Ventilation of emotions This may help, provided there is someone to whom emotions can reasonably be expressed. A close, understanding confidant(e) would be the obvious choice but such a person may not be available when needed or the patient may fear that using the person in this way may ruin a valuable relationship. Feelings might also be ventilated through other means, such as punching a rubber object or large cushion.

Physical exercise Vigorous exercise helps some individuals discharge tension. This might include running, swimming, exercises—i.e., anything which is physically demanding. Specific physical strategies which have also been suggested include the repeated squeezing of a rubber ball until this produces considerable discomfort in the wrist and forearm (Rosen and Thomas 1984), and wearing a rubber band on the wrist which can be flicked to produce pain but no further damage; distraction may be the principal effect of such strategies.

Physical contact A somewhat controversial approach suggested by Graff and Mallin (1967) is for the individual to seek out physical contact with another person. Some patients will have a partner or close friend who can provide physical contact when tension is becoming apparent. This strategy might be applied, with care, in the in-patient setting where physical comforting between nurses and patients is common, but should be restricted to staff of the same sex as the patient.

Medication Benzodiazepines and other minor tranquillizers appear to be contraindicated in the immediate prevention of cutting, being largely ineffective, possibly causing a paradoxical increase in aggression (Cowdry and Gardner 1988), although this risk has probably been exaggerated (Dietch and Jennings 1988), and having the risk of overdose and dependence. Neuroleptics may have a limited role—chlorpromazine, one of the other sedating phenothiazines, or haloperidol occasionally being of value.

Helping with underlying problems

As already indicated, self-cutters often display a range of psychological difficulties. Major problems which may require attention in most cases are low self-esteem, mood disturbances, and communication difficulties. Pursuing the meaning of cutting itself in a psychodynamic sense is not generally a helpful strategy.

Vulnerable self-esteem This problem is probably best addressed either through cognitive therapy or psychotherapy. Beck and colleagues have described how dysfunctional assumptions (basic beliefs that predispose the person to depression or other emotional disorders) may be identified and modified (Beck *et al.* 1979, Chapter 12) in the course of cognitive therapy. While there has been no systematic evaluation of this approach with self-cutters, clinical experience suggests it is worthwhile. There is evidence that cognitive therapy may be more beneficial than antidepressant treatment in improving low 'self-concept' (and reducing hopelessness) in depressed patients (Rush *et al.* 1982).

Mood disturbances Cognitive therapy can also be applied to the mood swings which most self-cutters report. In this the therapist will need to help the patient identify the cognitions which appear to provoke the mood changes (see Fig. 3). Often it will become clear that while attempts to question such cognitions may be helpful, the key to altering the patients catastrophic thinking depends on helping her understand how her dysfunctional assumptions lead to depressive cognitions when she experiences certain key events.

Prolonged mood disturbances, especially where these are associated with other cognitive or somatic symptoms of depression, may require specific treatment, either through psychological therapy (Fennell 1989) or drug treatment or both (the use of medication is discussed below).

Communication problems These may be addressed either in individual therapy, including the use of role play and role reversal, or group therapy, preferably with other patients with similar difficulties, although not necessarily self-cutting. Assertiveness training may also be helpful in this regard.

The *physical scars* resulting from repeated self-cutting can pose further problems for the patient because of the difficulty of explaining them to other people and the not unreasonable belief that self-cutting is likely to provoke revulsion in others. This awareness can further add to the patient's self-hatred. The therapist can try to help the patient develop a strategy for explaining the scars. Excusing them on the basis of an accident is only likely to be plausible when they are few and in certain sites (e.g., the back of the arm). Where there are multiple wrist scars it is probably best if the strategy includes an admission of self-cutting at a particularly difficult time previously in the patient's life and an explanation that the problems have long since resolved. Role play might help the patient develop confidence in such an explanation.

If psychological treatment is to address fully the extensive problems faced by many patients who cut themselves it cannot be brief. At least six months treatment may be required. Compliance, however, will often be a problem, especially in patients with major personality problems.

Medication

The main uses of medication in self-cutters are for preventing mood swings, reducing tension and aggressive feelings, alleviating more pro-tracted mood disturbances, and buffering the patient against the emotional impact of life events and other stresses.

As discussed in Chapter 9 (p. 107), positive effects on dysphoric mood states have been reported with low doses of neuroleptics, especially halo-peridol and trifluperazine, and with the monoamine oxidase inhibitor tranylcypromine. The possible association between aggressive behaviour, suicidal acts, and reduced brain 5-HT function (p. 109) may explain why lithium carbonate appears to be helpful in reducing aggression and self-mutilating behaviour in some patients with severe personality problems (see review by Wickham and Reid 1987). Likewise, carbamazepine has been reported to reduce episodes of behaviour disturbances, including suicidal acts, in patients with borderline personality disorders (Cowdry and Gardner 1988; and see p. 111). While so far there have been no controlled trials of such medication specifically in self-cutters, these drugs may warrant a trial in severe cases. However, compliance may be a problem and the risk of overdose must be carefully assessed.

It has been proposed that self-injurious behaviour may become self-reinforcing because calming and other pleasant effects result from release of endogenous opiates, specifically beta-endorphin. Although case reports have suggested that the opiate antagonist naloxone may attenuate or elimate self-injurious behaviour in mentally retarded patients, there is little evidence to date that naloxone may help self-cutters (Sandman *et al.* 1987).

Where there is evidence of depressive illness the usual treatments will be applicable. The potentially more activating antidepressants (e.g., imipra-mine) are probably best avoided since these may fuel the agitation that often leads to self-cutting. In the light of the possible abnormalities of serotonergic function noted above, antidepressants with specific effects on 5-HT pathways are probably worth trying (Primeau and Fontaine 1987). However, the significant risk of self-poisoning in these patients must again be borne in mind.

As discussed more fully in the next chapter (p. 111), in a placebo-controlled study of patients with a history of repeated suicide attempts it was found that depot flupenthixol decanoate (20 mg IM four weekly) reduced the frequency of further episodes (Montgomery and Montgomery 1982). While this study requires replication, it suggests an approach worth trying in severe cases, especially where other measures are unsuccessful or not acceptacle to the patient. A possible explanation for the success of this approach, if confirmed, could be that the neuroleptic medication has a protective effect against stresses similar to its function in schizophrenia where the patient lives in an emotionally stressful family environment.

Management in the in-patient setting

Admission to a psychiatric unit is occasionally necessary, especially if cutting is becoming increasingly frequent, there is thought to be a significant risk of suicide and/or the patient clearly has a severe depressive disorder. Self-cutting in hospital in-patient units is likely to be difficult to manage, but the following guidelines may minimize the problems and prevent the behaviour getting out of hand.

There should be a clear policy for the individual patient, with preferably one member of the nursing staff taking responsibility for the policy and any major decisions about changes in management. A team meeting may help increase other staff's understanding of the policy, and establish that they are happy to cooperate with it. Any restrictions on the patient should be clearly defined. Excessive restrictions are usually unhelpful, except for a brief initial phase, because the patient will often find a means of cutting in spite of them and they can become a major focus of dispute. However, limits, including what will happen if there are any further episodes of cutting, should be clearly established with the patient. A written contract, which should include what help will be offered as well as restrictions and limits, can establish a clear policy and is found helpful by some patients.

While self-cutting patients often welcome considerable attention from staff, this is best organized on the basis of regular therapeutic sessions, preferably with the same member of staff. An open acceptance of the patient's problems is essential. Treatment might focus firstly on methods of tension reduction, along the lines suggested earlier. The availability of exercise (e.g., a gym) or other means of discharging energy and aggression (e.g., a punch-bag) can be helpful. The patient might be encouraged to approach particular staff members whenever she feels tension, to talk about what has provoked it, and to try and combat it. Relaxation techniques (see above) may help, but initially are often more effective when conducted with the therapist than alone.

Subsequent therapeutic measures should be directed towards other problems, along the lines already discussed. If cutting does occur, physical treatment should be carried out with minimal fuss, certainly avoiding the situation where attention is contingent on cutting. The clinician may have to tolerate considerable frustration and anxiety over a long period before improvement occurs and may benefit from support from other members of the clinical team. In-patient admission should generally be kept as brief as possible; often it is only necessary to cover a period of crisis.

When an 'epidemic' of self-cutting becomes apparent in an in-patient unit, it is essential that a clear management policy be established. For example, discussion of the behaviour at community meetings is recommended. This might address the motivational aspects of the behaviour, ways help can be sought by other means, etc. Sometimes one, possibly two patients appear to be central to the epidemic and their removal from the unit may be necessary.

Clinical guidelines

Self-cutting clearly poses a major therapeutic challenge. The following is a summary of treatment strategies which may help prevent it.

(1) Help should initially be directed towards assisting the patient to try to gain some immediate control over cutting. A careful behavioural analysis should be carried out, following which appropriate strategies may be identified. These include controlling aggravating factors and utilizing factors which are known to prevent or delay cutting, helping the patient to recognize and reappraise the cognitive processes which underly rising feelings of tension and aggression, relaxation, ventilation of emotions, physical exercise, and physical contact. Medication has only a minor role in the immediate prevention of cutting. Benzodiazepines should be avoided.

(2) Cognitive therapy may help the patient understand and overcome the fragile self-esteem and negative reactions to seemingly minor stresses which often cause the rapid mood changes so characteristic of self-cutters. Individual or group therapy can help with communication and assertion difficulties. It may also be necessary to help the patient explain the scars to others.

(3) Medication has a role in management in some cases. Low-dose neuroleptics and monoamine oxidase inhibitors may counter dysphoric mood swings. In severe cases, a trial of lithium carbonate or carbamazepine is worthwhile. When an antidepressant is indicated, 'stimulating' drugs should be avoided. Antidepressants with specific serotonergic activity may be worth a trial. Special care in prescribing is necessary in these patients because of the risk of overdose. Compliance is often poor. Relatively low-dose depot flupenthixol may help patients who are especially sensitive to stress, where other measures are unsuccessful or unacceptable.

(4) Clear individual management policies are essential for self-cutters in the in-patient setting, and the clinical team should be vigilant for any signs that the behaviour is spreading to other patients.

(5) An understanding yet not overly sympathetic attitude on the part of the clinician is probably most helpful to the patient. The therapist may need considerable support from other team members since working with self-cutters can be extremely stressful.

Not much is known about long-term outcome. In one study of a small group of wrist-slashers, it was found that while a few ended their lives by suicide, more than half were well or had considerably improved five to six years later (Nelson and Grunebaum 1971). This finding, together with the fact that one rarely encounters self-cutters beyond young adulthood, suggests that a relatively optimistic attitude can be adopted as far as the longer term is concerned, even if the self-cutter currently seems trapped in an inescapable cycle of destructive behaviour.

Acknowledgements

I wish to thank David Clark, Joan Kirk, and Linette Whitehead for their helpful suggestions during the preparation of this chapter.

References

American Psychiatric Association (1987). *Diagnostic and statistical manual of mental disorders.* Third edition, revised. American Psychiatric Association, Washington DC.

Beck, A. T., Rush, A. J., Shaw, B. F., and Emery, G. (1979). *Cognitive therapy of depression.* Guilford, New York.

Cowdry, R. W. and Gardner, D. L. (1988). Pharmacotherapy of borderline personality disorder. *Archives of General Psychiatry,* **45,** 111–19.

Dietch, J. T. and Jennings, R. K. (1988). Aggressive dyscontrol in patients treated with benzodiazepines. *Journal of Clinical Psychiatry,* **49,** 184–8.

Favazza, A. R. (1987). *Bodies under siege.* Johns Hopkins University Press, Baltimore.

Fennell, M. (1989). Depression. In *Cognitive behaviour therapy for psychiatric problems* (ed. K. Hawton, P. M. Salkovskis, J. Kirk, and D. M. Clark), pp. 169–234. Oxford University Press.

Gardner, A. R. and Gardner, A. J. (1975). Self-mutilation, obsessionality and narcissism. *British Journal of Psychiatry,* **127,** 127–32.

Graff, H. and Mallin, R. (1967). The syndrome of the wrist-cutter. *American Journal of Psychiatry,* **124,** 36–42.

Grunebaum, H. U. and Klerman, G. L. (1967). Wrist-slashing. *American Journal of Psychiatry,* **124,** 36–41.

Hawton, K. (1978). Deliberate self-poisoning and self-injury in the psychiatric hospital. *British Journal of Medical Psychology,* **51,** 253–9.

Hawton, K. and Catalan, J. (1987). *Attempted suicide: a practical guide to its nature and management,* Second edition. Oxford University Press.

Montgomery, S. A. and Montgomery, D. (1982). Pharmacological prevention of suicidal behaviour. *Journal of Affective Disorders,* **4,** 291–8.

Nelson, S. H. and Grunebaum, H. (1971). A follow-up study of wrist-slashers. *American Journal of Psychiatry,* **127,** 1345–9.

Ost, L. G. (1987). Applied relaxation: description of coping technique and review of controlled studies. *Behaviour Research and Therapy,* **25,** 397–410.

Pattison, E. M. and Kahan, J. (1983). The deliberate self-harm syndrome. *American Journal of Psychiatry*, **140**, 867–12.

Primeau, F. and Fontaine, R. (1987). Obsessive disorder with self-mutilation; a subgroup responsive to pharmacotherapy. *Canadian Journal of Psychiatry*, **32**, 699–700.

Rosen, L. W. and Thomas, M. A. (1984). Treatment techniques for chronic wrist cutters. *Journal of Behaviour Therapy and Experimental Psychiatry*, **141**, 520–5.

Rosenthal, R. J., Rinzler, C., Wallsh, R., and Klausner, E. (1972). Wrist-cutting syndrome: the meaning of a gesture. *American Journal of Psychiatry*, **128**, 1363–8.

Rush, A. J., Beck, A. T., Kovacs, M., Weissenburger, J., and Hollon, S. D. (1982). Comparison of the effects of cognitive therapy and pharmacotherapy on hopelessness and self-concept. *American Journal of Psychiatry*, **139**, 862–6.

Sandman, C. A., Barron, J. L., Crinella, F. M., and Donnelly, J. F. (1987). Influences of naloxone on brain and behavior of a self-injurious woman. *Biological Psychiatry*, **22**, 899–906.

Simpson, M. A. (1975). The phenomenology of self-mutilation in a general hospital setting. *Canadian Psychiatric Association Journal*, **20**, 429–33.

Simpson, M. A. (1976). Self-mutilation and suicide. In *Suicidology: contemporary developments* (ed E. S. Schneidman), pp. 281–315. Grune and Stratton, New York.

Weissman, M. M. (1975). Wrist-cutting: relationship between clinical observations and epidemiological findings. *Archives of General Psychiatry*, **32**, 1166–71.

Wickham, F. A. and Reed, J. V. (1987). Lithium for the control of aggressive and self-mutilating behaviour. *International Clinical Psychopharmacology*, **2**, 181–90.

0

Personality disorders: are drugs useful?

PHILIP J. COWEN

The problem

Theory and practice

'Drugs have little part to play in the management of personality disorders' (Gelder *et al.* 1989, p. 149). This axiom, derived from a highly regarded textbook of clinical psychiatry, is undoubtedly a widely held view. For if personality disorders are 'deeply ingrained maladaptive patterns of behaviour' (*International Classification of Diseases*, Ninth Revision (ICD-9), WHO 1978), what role could there be for drug treatment except in the temporary alleviation of psychological distress? Even this palliation might be undesirable if it inhibited patients from adapting and modifying the unsuitable behaviour patterns that characterize their difficulties.

Despite this, it is apparent from clinical experience that psychotropic drugs are widely prescribed to patients with personality disorders. How can this gap between theory and practice be negotiated? I suggest that the same principle that guides drug treatment in other disorders should be applied, namely the assessment of the benefit-risk ratio of a particular drug treatment in a specific clinical situation. The fact that the biological basis of personality disorder is not understood need not deter us, because the same is essentially true of affective illness and schizophrenia. In these circumstances the test of any treatment is empirical; do the potential benefits outweigh the risks?

It is true that this judgement is peculiarly difficult to make in the drug treatment of personality disorders. There are two major reasons for this. First, the diagnosis of personality disorder is comparatively unreliable and its boundaries with other psychiatric disorders difficult to define. Second, partly because of the problem of diagnosis, there have been few well-controlled clinical trials of psychotropic medication. Such trials, however, are essential if the benefit of drug treatment is to be assessed accurately. More recently some well-designed studies have been published, the findings from which can be used to guide clinical practice in this difficult area.

Definition of syndromes

Diagnostic categories   The classification of personality disorders is still unsettled although operational diagnoses such as those employed in the

Diagnostic and Statistical Manual of Mental Disorders (DSM-III, American Psychiatric Association 1987) are of value in improving diagnostic reliability. Even in DSM-III the descriptions of personality disorders have overlapping features and some patients may meet criteria for more than one category. In addition, with some syndromes it is not yet possible to know whether the disorder is best viewed as a disturbance of personality or some other form of chronic psychiatric illness which has global effects on behaviour. For example, a patient with long-standing symptoms of affective disturbance may be classified as having an affective personality disorder in ICD-9 or a primary affective illness such as cyclothymia or dysthymia in DSM-III.

Classification by symptoms   Because of these problems in classification, it seems more useful to assess the effects of psychotropic medication on particular groups of symptoms rather than specific personality disorders. Even in other psychiatric disorders it appears that beneficial effects of psychotropic medication are not limited to particular diagnostic categories; for example tricyclic antidepressants are effective in the treatment of both mood and anxiety disorders. Accordingly this discussion will relate the effects of psychotropic drugs to particular symptoms commonly found in patients with personality disorders. Some symptoms associated with these disorders are more likely to result in the prescription of psychotropic drugs. These symptoms can be broadly divided into two categories; first, those associated with affective disturbance, and second, problems with impulsivity, hostility, and aggressive behaviour. The drug treatment of sexual disorders is not considered in this chapter.

Personality disorders and affective symptoms

Personality disorder and depressive illness

Recognition and treatment   Patients with personality disorders frequently experience episodes of major depression, and there is evidence that they are less likely than patients with uncomplicated depression to receive antidepressant drugs from psychiatrists. Even when antidepressant treatment is administered, however, the coexistence of major depression and personality disorder is associated with a less favourable outcome both in terms of increased suicide risk and persistent depressive symptoms at follow up.

Atypical depression   Personality factors may also modify the presentation of a depressive disorder and perhaps influence the response to different kinds of antidepressant treatment. It is suggested, for example, that patients with borderline and histrionic character traits are more likely to develop atypical depressive states which respond better to monoamine oxidase inhibitors than tricyclic antidepressants. These patients commonly show features of 'hysteroid dysphoria' such as reactively depressed mood, excessive interpersonal sensitivity, and histrionic behaviour, often triggered by loss of a romantic attachment. In a recent controlled investigation patients with this clinical presentation responded particularly well to phenelzine (up to 90 mg daily) but imipramine (up to 300 mg daily) was also reasonably effective (Liebowitz *et al.* 1988).

The authors of the latter study comment on the ability of phenelzine in some patients to reverse apparently long-standing symptoms of maladaptive behaviour and interpersonal difficulties. This finding shows the frequent difficulty of distinguishing long-standing personality disorders from character traits which have been exaggerated by illness. In any event, in many patients it will be best to reserve judgement on primacy of personality factors until an adequate trial of antidepressant treatment has been carried out (Liebowitz *et al.* 1988).

Personality disorder with dysphoria   Patients who meet DSM-III criteria for borderline personality disorder frequently experience intermittent but profound dysphoric mood swings which do not meet DSM-III criteria for major depression. Some recent double-blind trials have suggested that low doses of neuroleptics (for example, 7 mg daily of haloperidol) significantly attenuate these dysphoric states, perhaps more so than tricyclic antidepressants (Soloff *et al.* 1986). Positive effects on dysphoric symptoms have also been reported for tranylcypromine; but in the same investigation subjects who completed a full trial of trifluoperazine (8 mg daily) showed an equal subjective amelioration of depressive symptoms (Cowdry and Gardener 1988).

Chronic depression and personality disorder

Classification   It is widely accepted that major depressive disorders may run a chronic course in which neither response to treatment nor spontaneous remission occurs. There is debate, however, about the classification of less severe depressive states in which symptoms are also extremely persistent. In essence the difficulty is whether these disorders should be regarded as manifestations of a personality disturbance or an affective illness.

Dysthymia   Many of these patients will fulfil DSM-III criteria for dysthymia, which is classified as an affective disorder of less severity than major depression and with a predominance of cognitive rather than vegetative symptoms. In addition, for a diagnosis of dysthymia, the disorder must have been present for at least two years, though patients often report that depressive symptoms have been present for considerably longer. While the classification of chronic depression is still a matter of uncertainty, the major question for our purposes is whether drug treatment can reduce the depressive symptoms.

Drug treatment of dysthymia   In fact, few controlled studies are available to guide us. Often such patients have undergone psychotherapy, but they have rarely been thought suitable for antidepressant drug treatment. Recently, Kocsis and his colleagues (1988) described a double-blind trial of imipramine (up to 300 mg daily) in patients fulfilling DSM-III criteria for dysthymia. At the time of drug treatment, however, nearly all the patients also met criteria for major depression which had become superimposed on the long-standing dysthymic disorder (so-called 'double-depression').

At the end of the six week treatment period, clincial recovery had occurred in 59 per cent of patients completing the course of imipramine, but in only 13 per cent of those treated with placebo, a highly significant difference. It is noteworthy that to fulfil criteria for clinical recovery the patients had to be judged free of dysthymia as well as of major depression. This finding is rather remarkable when it is considered that the average duration of depressive symptoms in these patients was 19 years. Clearly in some subjects depressive symptoms thought to be inextricably bound up with maladaptive personality traits could be significantly ameliorated by a tricyclic antidepressant.

Several important questions remain to be answered. Will the antidepressant effect of imipramine in these subjects be maintained over a longer treatment period? Will the patients relapse when drug trcatment is withdrawn? Will a similar effect of imipramine be apparent in patients who only meet criteria for dysthymia and do not have an additional diagnosis of major depression?

Prediction of outcome   It seems clear that only a proportion of patients with dysthymia will respond to vigorous antidepressant medication? Can such patients be identified clinically before treatment? Based on open treatment studies, Links and Akiskal (1987) suggested that tricyclic antidepressants benefited dysthymic patients with a family history of affective illness. In addition, subjects who responded showed sleep EEG abnormalities characteristic of major depression. Patients whose depressive symptoms did not respond to antidepressants failed to demonstrate these clinical and biological features. Such patients often had first degree relatives with sociopathic traits and a history of substance abuse. The implication of these findings, which require prospective replication, is that patients with chronic depression who respond to antidepressant treatment have a form of affective illness, rather than a personality disorder.

Cyclothymia

Diagnosis   Similar difficulties in classification apply to patients who have mood swings that do not meet criteria for bipolar disorder. Often the mood swings are short-lived and not triggered by any obvious precipitant. Akiskal and co-workers (1977) distinguished a group of patients with a family history of bipolar disorder in whom the mood swings were accompanied by congruent behavioural changes (for example, decreased sleep and increased activity during spells of elevated mood). They suggested that such patients had a modified bipolar affective illness and contrasted them with subjects whose mood swings did not demonstrate congruent behavioural changes. In the latter group, rapid fluctuations in mood appeared to be secondary to a personality disorder (Akiskal *et al.* 1977).

Drug treatment of cyclothymia   The conclusion of this study (Akiskal *et al.* 1977) was that patients whose mood swings have an origin in affective illness may respond to pharmacological treatment, particularly lithium, but those in whom mood changes are secondary to personality disorder will not. However, controlled drug studies to test this suggestion are lacking. Indeed, in an early placebo-controlled study, Rifkin *et al.* (1972) found that lithium reduced mood swings in a group of patients described as having 'emotionally unstable character disorder'; however, the treatment period was only six weeks. Further controlled studies of lithium in patients with disruptive cyclothymic symptoms are needed.

Personality disorders and aggressive behaviour

Patients with certain kinds of personality disorders tend to exhibit persistently aggressive behaviours directed towards themselves and others. Frequently this aggression is of an impulsive nature which makes management difficult.

Diagnostic classification

In the DSM-III criteria, patients with clearly circumscribed episodes of explosive behaviour and no other psychiatric disorder are classified as having a disorder of impulse control, specifically intermittent explosive disorder; however, it is common for individuals who exhibit explosive outbursts to have evidence of other psychiatric disturbance, particularly antisocial or borderline personality disorder (Mattes 1986). A number of drug treatments have been suggested to be useful in the control of aggressive and impulsive behaviour. Such treatments may also be useful in reducing internally or externally directed feelings of hostility. While it is likely that all these symptoms describe rather different mental phenomena, they are usually considered together in studies of drug effects.

Neuropharmacology of impulsive and aggressive behaviour

5-hydroxytryptamine (5-HT) deficiency   For some time it has been believed that depressed patients, particularly those who exhibit suicidal behaviour, may have low levels of the 5-HT metabolite, 5-hydroxyindoleacetic acid

(5-HIAA), in cerebrospinal fluid (CSF) (see Virkunnen *et al*. 1987). While this association is still a matter of debate, there does seem to be a link between low CSF 5-HIAA and a tendency to behave in a violent and impulsive way. Thus reduced CSF 5-HIAA has been reported in naval recruits with a history of seriously aggressive behaviour; in arsonists; and in convicts who committed violent crimes impulsively. Interestingly, convicts who had carried out a premeditated crime of violence had normal CSF 5-HIAA levels (see Virkunnen *et al*. 1987).

Studies in animals suggest that lowered brain 5-HT neurotransmission may be associated with increased aggression and impulsivity. Taken together these findings raise the possibility that in some subjects reduced brain 5-HT function may be causally linked to impulsive and destructive acts. It is therefore of interest that some of the treatments thought useful in the management of aggressive behaviour such as lithium and carbamazepine do appear to increase brain 5-HT function (Virkunnen *et al*. 1987). Further studies are needed to confirm and identify the possible abnormality in brain 5-HT function in impulsive and violent offenders. Conceivably this could result in more rational treatments becoming available to ameliorate both the 5-HT abnormality and the aggressive behaviour.

Role of alcohol abuse   A factor complicating the above formulation is that many of the impulsive and violent subjects that have been studied have a history of alcohol abuse. Indeed, the acts which led to their convictions were usually committed under the influence of alcohol. Might low CSF 5-HIAA therefore in fact be a predisposing factor to the development of alcoholism? Alternatively, could reduced CSF 5-HIAA be a consequence of long-term alcohol abuse? These questions have yet to be definitively answered. Virkunnen *et al* (1987) have attempted to resolve the issue by suggesting that since acute administration of alcohol increases 5-HT neuro-transmission, subjects with low CSF 5-HIAA might drink to 'self-medicate', thereby relieving associated symptoms of dysphoria and tension. However, over longer periods of ingestion, alcohol may actually deplete brain 5-HT, leading to a vicious cycle of worsening alcohol abuse and increasing impulsivity. This intriguing hypothesis deserves further study.

Drugs used in the management of aggressive behaviour

Lithium   From controlled clinical trials it appears that lithium is the drug most established as effective in reducing episodes of aggression in patients with personality disorders. For example, in impulsively violent institu-tionalized offenders lithium markedly reduced serious incidents of violence (Sheard *et al*. 1976). While this work has some methodological limitations, a similar effect was recently reported in patients with mental handicap, where again the major change was a reduction of more serious aggressive behaviours (Craft *et al*. 1987).

In these studies the major criterion for the effectiveness of lithium was a reduction in frequency of aggressive incidents. There is no systematic data on whether lithium reduced feelings of hostility. Subjects reported that their anger, though still easily aroused, did not escalate in the usual way, perhaps suggesting that lithium may reduce impulsivity. However, patients do not generally report feeling subjectively better which could lead to problems in compliance.

Carbamazepine Carbamazepine is being increasingly used in the prevention of aggressive behaviour. While it was previously believed that the antiagressive effect of carbamazepine might be confined to patients with ECG abnormalities, this does not in fact seem to be the case (Mattes 1986). In a recent controlled crossover study carbamazepine significantly reduced episodes of behavioural dyscontrol, including suicidal acts, in borderline patients (Cowdry and Gardener 1988). Clinically, carbamazepine was thought to cause an increase in 'reflective delay' which seems to resemble the reported effect of lithium in impulsive subjects (see above). Also similar to lithium was the lack of subjective improvement reported by patients receiving carbamazepine.

Neuroleptics Neuroleptic drugs are widely used to control aggressive behaviour in patients with personality disorders, but until recently there were very few satisfactory double-blind trials of their efficacy. There are now, however, some well-designed studies that have examined the effects of low-dose neuroleptic treatment in patients with borderline personality disorder. In one investigation (Cowdry and Gardner 1988) the effect of neuroleptic treatment (8 mg trifluoperazine daily) was largely confined to subjective improvement in mood; in another, however (Soloff *et al.* 1986), haloperidol (7 mg daily) not only improved mood but also reduced hostility and interpersonal sensitivity. In addition there were reductions in scores measuring paranoid ideation and schizotypal thinking. The possible efficacy of neuroleptic drugs in borderline personality disorder is of interest in view of the finding that patients with borderline personality and schizophrenia share a common abnormality in auditory evoked potentials (Kutchner *et al.* 1986).

Neuroleptics and suicidal behaviour Neuroleptics may also reduce suicidal behaviour in patients with personality disorders. A study by Montgomery and Montgomery (1982) assessed the incidence of suicidal behaviour in a group of patients with a DSM-III diagnosis of personality disorder (mainly borderline or histrionic) who had made repeated attempts at suicide. In a six month double-blind study, depixol, 20 mg every four weeks, significantly reduced the incidence of suicidal behaviour in the latter half of the trial. This effect was not apparent in a parallel study where the antidepressant drug, mianserin, was employed. This investigation adds to the evidence that neuroleptic drugs in low dosage can reduce behavioural dyscontrol in patients with borderline personality disorder. However, Barraclough and colleagues (1974) have warned of a possible increased risk of suicidal behaviour in depressed patients treated with neuroleptics. Clearly, major depression should be excluded before depot neuroleptic treatment is started in patients with personality disorders.

Other drugs   The effects of benzodiazepines on aggression and hostility are complex and there are very few trials in patients with personality disorders. While it seems likely that benzodiazepines can reduce aggression in some subjects, behavioural disinhibition may also occur (Cowdry and Gardner 1988). This, together with the risk of dependence, is probably sufficient to discourage their use in the management of aggression in all but exceptional circumstances.

Recent open studies have suggested efficacy of propanolol in patients with explosive outbursts associated with organic brain damage (Mattes 1986). Whether this approach would be useful in patients with personality disorders is uncertain. Similar comments apply to the use of the antidepressant drug, trazodone, in patients with aggressive behaviour and organic brain damage (Pinner and Rich 1988).

Disadvantages of drug treatment

Unwanted effects   From the foregoing it will be apparent that certain drugs are efficacious in relieving some of the symptoms associated with personality disorder. However, all the drug treatments described have unwanted effects, some potentially hazardous. These need to be considered before any drug treatment is recommended to a patient (Table 8).

Overdose, dependence, and compliance   Drug treatment in patients with personality disorders may be associated with additional difficulties. With any patient who demonstrates impulsive behaviour, especially in the setting of dysphoric mood swings, the risk of overdose is significant. Patients with personality disorders are also likely to be less compliant with medication. This may apply especially to drugs such as lithium and carbamazepine where reductions in behavioural dyscontrol may not be associated with subjective improvement as far as the patient is concerned. With any drug treatment the risk of dependence must be considered; this is particularly likely to be a problem with benzodiazepines but may also be seen with tranylcypromine, which has stimulant properties. Related to this problem is the patient who continues to abuse alcohol. Clearly the possibility of an interaction between excessive alcohol intake and psychotropic medication must be considered.

Table 8 *Some adverse effects of psychotropic drugs used in the treatment of personality disorder*

Tricyclic antidepressants	Anticholinergic effects; psychomotor impairment; postural hypotension; lowered seizure threshold; impaired sexual function; high toxicity in overdose
Monoamine oxidase inhibitors	Postural hypotension; hypertensive reactions with some drugs and tyramine containing foods; insomnia; agitation; hypomania
Neuroleptics	Psychomotor impairment; acute extra-pyramidal movement disorders; lowered seizure threshold; tardive dyskinesia (with prolonged treatment)
Lithium	Tremor; thirst; polyuria; drowsiness; nausea; mental slowness; hypothyroidism; high toxicity in overdose
Carbamazepine	Ataxia; nausea; visual disturbance; drowsiness; skin rash; low plasma sodium; leucopaenia (rarely agranulocytosis); hepatitis (rare)

Duration of treatment   Another issue that should be mentioned is that nearly all the drug trials described here were of short-term duration, that is drug treatment was administered for weeks rather than months. Thus it is not certain that benefits reported from these studies would persist even if treatment was continued. Even more uncertain is whether improvements will be sustained when drug treatment is withdrawn. Indeed, unless the patient has learned new ways of dealing with difficulties, it would seem likely that cessation of drug treatment will lead to a recrudescence of symptoms. Further information on this point is needed.

Locus of control   Related to this is the likelihood that the most effective way of reducing aggressive behaviour is to improve the internal controls of the person concerned. It seems doubtful that drug treatment will increase feelings of personal responsibility for behaviour; indeed the probability will be that patients will see drug therapy as another external control which further confirms an inability to manage their own aggressive impulses.

Clinical guidelines

1. *Need for clinical judgement*

The indications for drug treatment in patients with personality disorders seem to be most clear when there is associated affective illness. In other situations, the decision to prescribe psychotropic medication depends largely on clinical judgement; thus differences in individual practice are likely to occur.

2. *Depression*

Patients with personality disorders who have episodes of major depression should be treated with antidepressant drugs. There is some evidence that patients with features of atypical depression may respond well to mono-amine oxidase inhibitors. In such cases, apparent features of a personality disorder, particularly histrionic or borderline characteristics, may remit with successful treatment. It also seems reasonable to offer a trial of medication to patients with chronic affective symptoms, not necessarily meeting DSM-III criteria for major depression or bipolar illness, especially where a family history of affective disorder is apparent. Imipramine and monamine oxidase inhibitors have been reported as effective in some patients with apparently 'characterological' depression. Lithium is worth trying in patients with cyclothymic symptoms, particularly if these are associated with mood-congruent behavioural changes.

3. *Dysphoria and dyscontrol in borderline personality*

In patients who meet DSM-III criteria for borderline personality disorder, neuroleptic drugs in modest dosage may be effective in reducing both dysphoric symptoms and episodes of behavioural dyscontrol. The possible adverse effects of neuroleptics (Table 8) suggest that such drugs should be given for a period of a few weeks as an adjunct to other forms of treatment. However, patients in whom repeated episodes of self-harm are associated with borderline personality features may merit longer-term treatment with a neuroleptic drug. The use of a depot preparation in these subjects is worth considering because compliance is increased and overdosage with prescribed medication cannot occur. The use of a monoamine oxidase in-hibitor in borderline patients who do not have an additional major affec-tive disorder is more difficult to justify at present because of the need for close compliance with treatment together with strict adherence to dietary restrictions. If further trials suggest that monoamine oxidase inhibitors are particularly useful in such patients, this view should be reconsidered.

4. *Treatment of aggressive behaviour*

Both lithium and carbamazepine may be effective in reducing impulsive and aggressive behaviour in patients with diverse kinds of personality disorder. At present the weight of evidence from controlled trials would favour lithium but some patients will tolerate carbamazepine better. With both drugs, routine blood testing to measure plasma levels is carried out, though probably this is less necessary with carbamazepine where there is little evidence to link psychotropic effects with an identified plasma level. Treatment with lithium or carbamazepine, if effective, may have to be maintained for a long period of time. In addition, patients may not themselves note subjective improvement. These drawbacks indicate that drug treatments of aggression should, initially at least, be used for short periods of time to supplement other therapeutic approaches.

References

Akiskal, H. S., Djenderedjian, A. H., Rosenthal, R. H., and Munir, K. K. (1977). Cyclothymic disorder: validating criteria for inclusion in the bipolar affective group. *American Journal of Psychiatry*, **134**, 1227–33.

American Psychiatric Association (1987). *Diagnostic and statistical manual of mental disorders*. Third edition, revised. American Psychiatric Association, Washington, D.C.

Barraclough, B., Bunch, J., Nelson, B., and Sainsbury, P. (1974). A hundred cases of suicide: clinical aspects. *British Journal of Psychiatry*, **125**, 355–73.

Cowdry, R. W., and Gardner, D. L. (1988). Pharmacotherapy of borderline personality disorder. *Archives of General Psychiatry*, **45**, 111–19.

Craft, M., Ismail, I. A., Krishnamurti, D., Mathews, D., Regan, A., Seth, R. V., and North, P. M. (1987). Lithium in the treatment of aggression in mentally handicapped patients: a double-blind trial. *British Journal of Psychiatry*, **150**, 685–989.

Gelder, M. G., Gath, D. H., and Mayou, R. (1969). *Oxford textbook of psychiatry*, Second edition, (pp. 149). Oxford University Press.

Kocsis, J. H., Frances, A. J., Voss, C., Mann, J.J., Mason, B. J., and Sweeney J. (1988). Imipramine treatment for chronic depression. *Archives of General Psychiatry*, **45**, 253–7.

Kutchner, S. P., Blackwood, D. H. R., St Clair, D., Gaskell, D. F., and Muir, W. J. (1986). Auditory P300 in borderline personality disorder and schizophrenia. *Archives of General Psychiatry*, **44**, 645–50.

Liebowitz, M. R., *et al.* (1988). Antidepressant specificity in atypical depression. *Archives of General Psychiatry*, **45**, 129–37.

Links, P. S. and Akiskal, H. S. (1987). Chronic and intractable depressions: terminology, classification, and description of subtypes. In *Treating resistant depression* (eds. J. Zohar and R. Belmaker), pp. 1–22. PMA Publishing Corporation, New York.

Mattes, J. A. (1986). Psychopharmacology of temper outbursts. *Journal of Mental and Nervous Disorders*, **174**, 464–70.

Montgomery, S.A. and Montgomery, D. (1982). Pharmacological prevention of suicidal behaviour. *Journal of Affective Disorders*, **4**, 291–8.

Pinner, E. and Rich, C. L. (1988) Effects of trazodone on aggressive behaviour in seven patients with organic mental disorders. *American Journal of Psychiatry*, **145**, 1295–6.

Rifkin, A., Quitlin, F., Carillo, C., Blumberg, A., and Klein, D. (1972). Lithium carbonate in emotionally unstable character disorder. *Archives of General Psychiatry*, **27**, 519–23.

Sheard, M. H., Marini, J. L., Bridges, C. I., and Wagner, E. (1976). The effect of lithium on impulsive aggressive behaviour in man. *American Journal of Psychiatry* **133**, 1409–13.

Soloff, P. H., George, A., Nathan, S., Schulz, P. M., Ulrich, R. F., and Perel, J. M. (1986). Progress in pharmacotherapy of borderline disorders. *Archives of General Psychiatry*, **43**, 691–7.

Virkunnen, M., Nuutila, A., Goodwin, F. K., and Linnoila, M. (1987). Cerebrospinal fluid monoamine metabolite levels in male arsonists. *Archives of General Psychiatry*, **44**, 241–7.

WHO (World Health Organization) (1978). *Mental disorders: glossary and guide to their classification in accordance with the ninth revision of the International Classification of Diseases*. World Health Organization, Geneva.

10

Dangerousness: which patients should we worry about?

JOHN R. HAMILTON and HENRIETTA BULLARD

Concepts and predictions

'Avoiding danger is no safer in the long run than outright exposure. The fearful are caught as often as the bold'—Helen Keller, *Let us have faith*, 1940

The question in the title of this chapter would no doubt be answered differently according to one's standpoint. The cynic might say 'all patients' or 'none of them' as there are no scientific measures to predict accurately dangerousness. The libertarian would say one should be concerned about the civil liberties of all offender patients detained in hospitals as only a minority of them will behave dangerously in the future. For the psychiatrist, however, this chapter will attempt to give some practical guidelines on the assessment of those patients who do cause concern.

Definitions

The Home Office and DHSS (1975) Committee on Mentally Abnormal Offenders defined dangerousness as 'a propensity to cause serious physical injury or lasting psychological harm'. The definition given by Scott (1977) makes no reference to psychological harm (which can be a consequence of some sexual offences and of verbal aggression): 'Dangerousness is an unpredictable and untreatable tendency to inflict or risk serious, irreversible injury or destruction.' Here destruction would presumably include the deliberate or reckless arsonist, but many would take issue with the assertion that the condition is untreatable, or indeed entirely unpredictable. For the purpose of this chapter the word 'dangerousness' means the risk of inflicting serious violence on others, causing serious psychological harm, and damaging property where there is a risk of physical injury to others.

Dangerousness is not a measurement like height or intelligence: it is a judgement made on the basis of information available and, as such, is potentially invalid and unreliable. This judgement cannot be made without regard to the personal circumstances of the individual. One does not worry about the paedophile who is detained in a special hospital where he has no access to children, but there would be concern if on release he obtained employment as a caretaker at a school. The dangerousness of an individual, then, is not a constant quality but liable to change according to his relationship with others and the stability of his mental state.

Prediction studies

No one has ever devised a test which can satisfactorily predict violent behaviour by an individual. 'Success' rates claimed by psychiatrists in various studies (Hamilton 1982) range from 0.33 per cent to 40 per cent but in the latter study there was a false positive rate of 65 per cent. 'False positive' patients are those who it is predicted *will* be violent but turn out not to be, whereas false negatives are cases assessed as *not* being dangerous but later commit violent acts. In England, special hospitals are provided for patients of 'dangerous, violent, or criminal propensities' (Section 4, National Health Services Act 1977), but it is not known what proportion of the 1700 patients detained therein are false positives who would not offend on release. It is to be expected that those who have the power to initiate or authorize release—psychiatrists, the Home Office, and Mental Health Review Tribunals—will err on the side of caution. Psychiatrists are criticized for being over-cautious, detaining patients for too long in conditions of inappropriately high security. The evidence from research studies of patients released from high security hospitals in the USA suggests there may be two false positives for every true positive. Because of the difficulty in predicting when patients will recover from their mental disorders and when they will no longer be dangerous, legislation provides that patients can be detained indefinitely on treatment orders and hospital orders (sections 3 and 37 of the Mental Health Act 1983), and restriction orders (section 41) can be made without limit of time. Unless patients are to be detained forever, decisions have to be made as to the optimum time for discharge.

While psychiatrists are rightly criticized for their poor performance in predicting dangerousness, the nature of their work does require them to make judgements on the admission of patients to hospital, the level of security required, appropriate treatment, and the likelihood of future offending. Psychiatrists should use their training and experience to assist them in their clinical assessments. They need to know why patients behaved dangerously in the past and in what circumstances they are likely to repeat that behaviour in the future.

Dangerousness and mental disorder

Are psychiatric patients more dangerous than normal people? The answer of course depends on what is meant by 'psychiatric patients', who could be in-patients, out-patients, or those not yet recognized as having a mental disorder. A prediction of future dangerousness is usually made by examining the history of previous dangerous behaviour and relating this to changes in a number of parameters, including the mental state. A record of convictions may not reflect the totality of a patient's potential for dangerous behaviour, but is an important guideline. Monahan (1989) has described the problems inherent in research into reconvictions: for instance schizophrenics discharged from hospital who behave violently in the community are often not charged but taken to hospital. Patients predicted by hospital staff as likely to be imminently dangerous are unlikely to be released, making such predictions inaccessible to testing. Within hospitals, patients who are thought likely to become violent may be sedated, secluded, or transferred to more secure accommodation: the violence does not occur and the prediction is untested. The wide variation of follow-up periods in research studies, the lack of co-ordination in defining predictor and criterion variables, and other factors make replication and comparison difficult if not impossible.

One major study of the incidence of mental abnormality in offenders was conducted in West Germany by Hafner and Boker (1973) who reviewed all those convicted of serious violence in a ten-year period. They defined mental disorder as including schizophrenia, affective psychosis, organic psychosis, and mental handicap and found that only those suffering from schizophrenia were over-represented among violent offenders. Similarly, Taylor and Gunn (1984) found a higher prevalence of schizophrenics compared to the general population in remanded male prisoners charged with violent offences in England. Contrary findings have been published by others and much more research is needed to clarify whatever connection may exist between violence and mental disorder. Meantime Monahan (1981) has concluded that the best predictors of violence among the mentally disordered are the same factors (age, sex, social class, and history of previous violence) that are the best predictors among those who are not mentally disordered; and that the poorest predictors of violence among the mentally disordered are psychological factors such as diagnosis, severity of mental disorder, and personality traits.

There does not appear to be any clear evidence that violence is substantially higher among psychiatric patients in general, but this may not be quite so true for patients with schizophrenia, psychopathic disorder, and alcoholism. Even in these patients there are no statistics or tests which will help the psychiatrist predict violence with any degree of accuracy. All we can safely say is that the best guide to future behaviour is previous behaviour and the more offences previously committed by an individual the more likely it is that he or she will offend again (Monahan 1981).

Acquiring a knowledge base

'Mystery magnifies danger as fog the sun'— Charles Coleb Cotton, *Lacon*, 1925.

Given that assessments of dangerousness are still expected and required of psychiatrists and that this will essentially be a clinical task, a full history must be taken together with a thorough examination of the mental state. It cannot be emphasized too strongly how important these matters are: 'If you don't ask the right questions you won't get the right answers.' Patients are often relieved to be able to discuss their fantasies and compulsive urges and to admit their fear of losing control of their behaviour.

Domestic violence

Just as psychiatrists are not afraid to ask depressed patients about suicidal ideation, aggressive patients should be asked about their violent thoughts and what they have done in practice. Spouses and others should be asked about domestic violence. Post *et al.* (1980) found that patients do not readily offer information about the violence they have experienced unless asked directly. In their study of psychiatric in-patients 48 per cent of females revealed they had been battered while 21 per cent admitted battering their husbands. Of male patients 27 per cent admitted battering their partners and 14 per cent said they had been battered. Gayford (1979) classified battered wives into various types including: the inadequate woman married to a man of similar characteristics; the competent wife married to her intellectual inferior who feels he has to assert himself; and the attention-seeking, provocative wife who enjoys the company of men. Husbands rarely ask for help, other than some middle-class men who fear losing their social position; they may be amenable to psychotherapy. A few men complain of violent rages and may need EEG investigation for possible underlying seizure disorder. Violence starting in later life may indicate loss of control associated with a dementing process.

Child abuse

Smith and Hanson (1974) found the majority of mothers of child abuse victims to be emotionally immature and many were depressed. One third of the fathers had a psychopathic personality. These parents tend to come from unhappy backgrounds and many were themselves battered as children. Social circumstances are often poor, with poverty, isolation, and unemployment common. Children most at risk are poor feeders, non-cuddlers, those who cry incessantly, are not amenable to toilet-training, are hyperkinetic, or particularly unresponsive: they may trigger off a violent response in a vulnerable parent. There is an increased incidence of physical or mental handicap in such children, which, in a mother with poor coping mechanisms and little support from her partner, can lead to scapegoating, the abused child becoming the displaced focus for all the family's ills and discontents.

Homicide

When children are killed, it is usually by their parents. D'Orban (1979) classified such mothers into battering mothers, mentally ill mothers, and smaller numbers in the categories of the 'Medea' (revenge) syndrome, mercy killing, and neonaticide. Half the 'battering' mothers had a previous history of such violence and in most the child's behaviour (often due to illness) was the immediate stimulus for the offence. These precipitants included persistent crying or screaming, feeding problems or vomiting, incontinence or 'messy' behaviour, or a child being emotionally unresponsive after a period of separation. Recognition of such triggers is important, not only in prevention but in treatment and prevention of a repetition. In most homicides in England and Wales there is a relationship between offender and victim. In 50 per cent the victim is a family member, spouse, cohabitee, or lover. In 25 per cent the victim is an acquaintance, and only 25 per cent are strangers. If the offender is mentally abnormal the victim is usually a member of his or her family, if mentally normal a stranger. In Scotland a stranger is more likely to be killed by a mentally normal individual and the mentally abnormal kill family members as often as strangers. Most killings therefore are domestic, and offender or victim or both may or may not be mentally abnormal. Many homicides arise from jealousy, quarrels, or loss of temper against a background of marital discord, sexual problems, and financial difficulties. Of adult women victims 40 per cent are killed by their husbands and a further 25 per cent by a relative or lover. Pathological jealousy account for 12 per cent of Broadmoor psychotic killers with diagnoses of schizophrenia, depression, and alcoholism. The wife's imagined lover is at equal risk.

Bluglass (1979) found half of spouse murder victims were alcoholic, psychotic, epileptic, or physically disabled and had played a significant part in their own deaths. Sons who commit matricide are usually schizophrenic and daughters who kill their mothers are nearly always psychotic. Fathers who are killed by their sons are usually overbearing, brutal, or alcoholic. In sexual homicide the victims are often prostitutes or deprived or neglected children and the murders may include bizarre sadistic mutilations. Sometimes the killing occurs in panic after a homosexual act, following rape, or the indecent assault of a child. If mentally abnormal, killers may suffer from schizophrenia, depression, mental handicap, or psychopathic disorder and rarely from epilepsy or organic brain disorder. Alcoholic abuse or drug intoxication may be associated factors.

Sex offenders

Incest Alongside child abuse, incest has recently become a topic of much public attention. Most cases which come before the courts are father–daughter incest. The father rarely has a major psychiatric illness but may well have personality disorder of inadequate or aggressive type. Faulk (1988) describes three types of father who commit incest. The first is an aggressive psychopath in his forties who abuses alcohol and is likely to have a history of violence, a poor work record, and who probably came from a broken home. His offences may involve more than one daughter while he concurrently has a sexual relationship with his wife or perhaps mistress. The second is the inadequate psychopath who may well be unemployed with his wife being the breadwinner and who seduces his daughter by exploiting her sympathy for his unsatisfactory social situation. The third group of fathers may commit incest when their wives are ill or away from home. Major psychiatric illness is however often found in the mother in cases of mother–son incest.

Exhibitionism   This is the commonest sex offence and may occur as a consequence of a lessening of inhibitions in early dementia or immaturity associated with mental handicap. Some men may expose themselves at times of stress, there often being a compulsive nature to their acts, the relief of tension being followed by guilt. It would appear that such individuals are not likely to proceed to more serious sex offences and are indeed good candidates for treatment. On the other hand there are exhibitionists with seriously disordered personalities who masturbate during their exposing: some of them may go on to commit serious aggressive offences.

Paedophilia   This likewise may sometimes be a consequence of organic brain disease or as an 'out of character' reaction to stress and is not uncommon in mentally handicapped men where the physical age of the victim is in line with the offender's mental age. The more 'dyed in the wool' paedophile will probably have been a victim of the same offence as a child. Many such men find employment centred around children—as a teacher, clergyman, scoutmaster, or youth club worker—are sincerely affectionate towards children and are impotent in adult sexual relationships. Ingram (1979) comments 'It is easy to see how men deprived of love as children should find in the unhappy victims objects of a deep love, and how the deprived children would cling to such men for the love they had to offer.' The more dangerous paedophile is one who has little capacity for affection for anyone, including his victims, whom he will bribe, threaten, or force to engage in sexual activities. The victims may be male or female, prepubertal or pubertal. The psychological damage to the victims will be greater when they are less than willing sexual partners. It is generally agreed that someone who rapes a child is likely to be mentally abnormal. Increasing age does not necessarily diminish the libido of a paedophile. The more previous convictions an offender has, the higher the risk of repetition.

Rape   This subject is discussed with much emotion, misinformation, and mythology. Criminal statistics show that 10 per cent of those convicted are under the age of 17 and 40 per cent aged 17 to 21. Multiple rape is much more common than many people realize: in 40 per cent of cases there is more than one offender. Rape is of course an aggressive offence which involves not only forced sexual intercourse but is often accompanied by degrading sexual acts and threatened or actual violence. Classifications of rapists are often of an arbitrary nature and may be based on the behaviour of the offender, his psychological characteristics, or the type of victim. In clinical practice the offender may be found to have a personality disorder, mental handicap, mental illness, or none of these. In most cases the offender is not mentally abnormal, but will be a young man who is sexually inexperienced, immature, and sexually aggressive. He may be reckless as to whether the victim was consenting. Rape rarely occurs as a result of mental illness. If it does, the offender may be suffering from schizophrenia, hypomania, dementia, or brain damage resulting in disinhibition. Mental handicap is associated with immature social and psychosexual development and rape may be a consequence of lack of recognition of generally accepted standards of sexual conduct or of overpowering sexual frustration. Personality disorder associated with rape may be of inadequate or aggressive type. The inhibited, socially inadequate individual may be prevented by anxiety from making contacts leading to a sexual relationship, instead engaging in indecent assaults (touching women's buttocks or breasts and running away) and progressing to rape. The more aggressive psychopath is likely to have previous convictions for violence or other offences and to abuse alcohol or drugs. In such cases the act of rape can be seen as another form of aggression and the individual may be impotent unless his victim resists him. Rapists with severe personality disorders are extremely difficult to treat, though if young, motivated, and with some degree of insight into their psychopathology they may, in the UK, be accepted for admission to one of the special hospitals or receive treatment in Grendon psychiatric prison.

An uncommon variant of the aggressive psychopath is the sadistic psychopath who hates women and who, besides inflicting violence on his victims, may proceed from humiliating, torturing, and defiling them to mutilating and killing them. Brittain (1970) described the syndrome of such sadistic murderers who interest themselves in perverted sexual matters, torture, Nazism, and the occult and who may watch 'video nasties' and horror films and collect pornography and weapons. MacCulloch *et al.* (1983) have described how sadistic murderers fantasize and rehearse their attacks before carrying them out.

Sex offenders have a reputation for recidivism but criminal statistics show that 90 per cent of men convicted for rape for the first time are not convicted for rape again. Soothill *et al.* (1976) found, however, that about 15 per cent are later convicted of a less serious sexual assault and a similar number for violent offences. From a 22-year follow-up study, Soothill and colleagues (1978) also showed that those acquitted of rape are as likely to be convicted later of rape as those convicted and that re-offending can take place after many years.

Arson

As with rape, several attempts have been made at classifying arsonists. Perhaps the most helpful is that of Faulk (1988), based on clinical experience. Individuals may come into more than one category and psychiatric disorder and alcoholism may be found in any of the subgroups.

In the first main group, the fire is a means to an end and psychiatrists will rarely be asked to examine those who have started a fire to cover up evidence of a crime or as part of an insurance fraud. Other motives may be revenge, anger, or jealousy committed by, for instance, a disgruntled or sacked employee at his place of work or a schoolchild to school buildings. Fire-setting can be a gang-activity in adolescents but the instigator may be seriously disturbed with concomitant truancy, running away from home, and neurotic traits. Mentally ill individuals may set fires as a 'cry for help' when seriously depressed, or as a result of delusions or auditory hallucinations in schizophrenia. These subgroups will probably have set relatively few fires: if there have been more than a few set by the same individual he (or she) may feel (and be) inadequate and enjoy the feeling of power or of being seen as a hero at the site of the conflagration.

In the second main group, the fire itself is the focus of interest. Rather like the compulsive exhibitionist there is a group of people who harbour irresistible impulses to set fires to relieve intolerable anxiety. They may previously have set fires of escalating severity.

Last, and least, are those whose motive is sexual excitement. Such cases do exist but are uncommon: the offender is likely to be an immature young man with considerable psychosexual pathology.

Assessment of arsonists should be aimed at elucidating the psychopathology, motives, and precipitants for the fire-raising and obtaining an account of previous similar behaviour and what the patients themselves say about their fantasies, desires, or impulses to set fires. Some arsonists appear oblivious to the danger any fire can cause to human life. The seriousness with which the law regards the offence is reflected in the possible sentence of life imprisonment for one who is convicted of the offence where he or she intends to endanger life or is reckless in this regard.

The most important study of recidivism in arsonists was by Soothill and Pope (1973) who, in a 20-year follow-up, found only 4 per cent set fires again though 10 per cent committed subsequent sexual or violent offences. Other research has shown that 20 per cent of those who had received a long prison sentence for arson committed the offences again (Sapsford *et al.* 1978). Overall the chance of a first offender repeating the offence is very low but the risk of repetition appears higher when the fire-raising is associated with mental illness, in those who have compulsive urges to set fires to relieve tension, and where the motives are sexual excitement or 'revenge' against society in general.

Assessment of dangerousness according to psychiatric diagnosis

A careful assessment of patients who either have been or might become dangerous is of the utmost importance. This should include the personal history, previous psychiatric and forensic history. The history of previous offences and dangerous or assaultive behaviour is particularly important and details must be obtained, even though this is sometimes a challenging task. It is also important to know the relationship between the dangerous behaviour and the illness and whether the patient was having treatment or, if not, when treatment or supervision was stopped. A comprehensive psychiatric history will help the doctor to formulate a profile of the patient and will assist him or her in making the correct diagnosis and instituting the most appropriate treatment. The mental state at the time of the violent incident or offence should be noted and a judgement reached as to the probable sequence of events which precipitated the violence. The diagnosis may be an important predictor of a patient's behaviour (p. 119) and the prognosis will largely depend on whether the illness can be treated and whether appropriate supervision can be provided for the patient in the community.

Schizophrenia

Many patients with acute schizophrenic illnesses present with disturbed behaviour. This may involve bizarre behaviour which is not in itself dangerous; or unexpected and unexplained aggressive behaviour either to property or people. Such behaviour usually responds rapidly to treatment and is not seen as dangerous. The patient with paranoid symptoms must be taken more seriously and an assessment made as to the likelihood of the patient acting on paranoid delusions. A middle-aged woman may complain for years that the neighbours are pumping gas under her floorboards and will never take action against them. On the other hand, a young male schizophrenic who believes his mother has stolen part of his brain may feel so persecuted that he kills her. The schizophrenic with a well-preserved personality can be just as dangerous as the schizophrenic who, in addition to persecutory delusional beliefs, has marked deterioration in his personality. How then do we assess an individual patient's potential for dangerousness? If any of the following circumstances are present, the doctor should be alert to the possibility that the risk of serious violence is substantial.

1. The patient has a history of assaults and has used a weapon.
2. The patient has expressed an intention to harm or kill and has a named victim or victims.
3. The patient lives alone with his or her parents and either has delusions concerning them or has made previous assaults on them. There is

evidence that where family violence and schizophrenia are associated, the mother is at the greatest risk.

4. Delusional jealousy has a poor prognosis and is sometimes seen in connection with a schizophrenic illness. The victim is most likely to be the wife but as noted earlier, the wife's supposed lover is also at risk. The poor prognosis associated with delusional jealousy is related to the difficulty of treating the illness. Patients frequently do not regain insight and years after an offence involving violence towards his wife, a patient will persist in his delusional belief concerning her infidelity.

Manic depressive psychosis

The manic patient may be irritable, highly aroused, and can present very serious management problems. However, except in very resistant cases the manic episode is short-lived and can be controlled with drugs. It is very rare for manic patients to be responsible for causing serious injuries or death. They do cause damage to property and strongly resent attempts to restrain their natural exuberance.

Depressive illness is commonly associated with suicide and 15 per cent of depressed patients succeed in committing suicide at some time. Homicide and suicide are closely related and where husbands kill their wives the diagnosis is predominantly one of depression. One-third of men charged with a domestic murder kill themselves; of the remaining two-thirds, half attempt suicide. Depressed men who kill their wives fall into two groups. First, those with a good pre-morbid personality and a good marriage, where the killing is altruistic. Such individuals may believe that they or their wives have fatal illnesses or for some other reason their spouse's life is not worth living. Secondly, those with dependent and possessive relationships with their wives whom they may or may not always treat well. Such marriages are often of long standing and violence is not a feature until the husband becomes depressed. A depressive illness can expose histrionic, inadequate, and dependent personality traits in a man, who may then believe that his wife has lost interest in him or has a lover. She may become irritated by his behaviour and decide to leave him. He then becomes agitated and in a state of despair makes an unpremeditated attack. On examination of a series of husbands who killed their wives there was evidence of previous assaultative behaviour before the fatal attack. It is very important to admit to hospital for treatment any man who has attempted to strangle his wife or made any other serious assault where there is no previous history of marital violence.

The following circumstances should alert the doctor to the possibility of dangerous behaviour in a depressed patient.

1. A woman who is suffering from post-natal depression and who attempts to harm or kill her child, even when she has not stated any intention to harm the child.

2. Depressed patients who try to kill themselves and their partners: this is particularly relevant in the elderly where the partners are dependent on one another for all their needs.

3. The depressed patient who expresses suicidal and homicidal preoccupations. Such patients may be a danger to their families, particularly men towards their wives.

Neuroses

Neurotic disorders are uncommonly associated with serious violence. Patients suffering from hysteria may exhibit very bizarre and sometimes dangerous behaviour during dissociative states. Such behaviour is invariably 'forgotten' by the patient, who when seen, characteristically behaves as if nothing had happened.

Personality disorders

These tend to be over-diagnosed; for example, antisocial behaviour in chronic schizophrenics or young delinquents may be seen as evidence of a personality disorder. Serious violence is usually a product of psychopathic disorder. Such disorders can be subdivided into aggressive and inadequate. All types of antisocial behaviour are associated with the aggressive psychopath, particularly unsolicited violence. The sexual psychopath is a separate entity and the individual may not exhibit any violent behaviour other than in a sexual context. It is important to question the sex offender about sexual fantasies and to establish whether there are any previous convictions for sexual offences. Sexual and other forms of sadism are commonly associated with the need to control others and an interest in controlling regimes such as fascism. The following factors may be helpful in assessing patients with psychopathic personality disorders.

1. Previous behaviour is a particularly good predictor of future behaviour.

2. Psychopathic disorders in women can be associated with serious violence such as arson. Self-mutilation and the taking of overdoses are often concomitants of violence in women.

3. An associated depressive illness accentuates the symptomatology associated with all personality disorders.

4. In assessing a patient's potential for future violence, it is important to establish whether the patient has experienced periods of stability when he or she has not exhibited violent behaviour. The personality-disordered patient is particularly sensitive to environmental influences.

Mental handicap

Some patients suffering from mental handicap do express their frustrations in violence to others and in inappropriate and sometimes violent sexual behaviour. The mentally handicapped patient may require experienced nursing in an institutional setting to prevent inappropriate sexual advances to children and other aggressive and disinhibited types of behaviour. Non-violent sex offences, including exhibitionism and indecent assault, are sometimes associated with the mentally handicapped patient, as is arson.

Organic brain disorders

The organic dementias are a rare cause of dangerous behaviour. The commonest type of referral to the forensic psychiatric service is the brain-damaged young male with a frontal lobe syndrome following a road traffic accident. The patient is generally disinhibited, but particularly sexually. He may make inappropriate sexual advances to women which include touching, stroking, and improper suggestions. In a very small minority of cases, the patient may demonstrate violent and assaultative behaviour.

Clinical guidelines

1. It has to be acknowledged that the prediction of dangerousness is extremely difficult. In assessing risk of dangerousness, a carefully taken psychiatric history and mental state examination are essential. The assessment should include direct enquiry about violent thoughts and previous violent behaviour.

2. The best predictor of future dangerous behaviour is previous behaviour of a similar nature. Diagnosis is a less useful predictor, except that schizophrenia, psychopathic personality disorder, and alcoholism are associated with increased risk of violence. The risk in those with personality disorders is increased by a concomitant depressive disorder.

3. The psychiatrist should be vigilant for possible organic disorders underlying violent behaviour. These include dementia, epilepsy, and frontal lobe damage.

4. In assessing the risk of particular offences, heed should be paid to the findings, reviewed in this chapter, of previous research indicating the common characteristics of the potential offender. Not only should the assessment concern the history, behaviour, and mental state of the potentially violent individual, but also the person's environment and the behaviour and condition of the potential victim.

References

Bluglass, R. S. (1979). The psychiatric assessment of homicide victims. *British Journal of Hospital Medicine*, **20**, 286–90.
Brittain, R. P. (1970). The sadistic murderer. *Medicine, Science and the Law*, **10**, 198–207.

D'Orban, P. T. (1979). Women who kill their children. *British Journal of Psychiatry*, **134**, 560–71.

Faulk, M. (1988) *Basic forensic psychiatry*. Blackwell, Oxford.

Gayford, J. J. (1979). Battered wives. *British Journal of Hospital Medicine*, **22**, 496–503.

Hafner, H. H. and Boker, W. (1973). Mentally disordered violent offenders. *Social Psychiatry*, **8**, 220–9.

Hamilton, J. R. (1982). A quick look at the problems. In *Dangerousness: psychiatric assessment and management* (ed. J. R. Hamilton and H. Freeman), pp. 1–4. Gaskell, London.

Home Office and Department of Health and Social Security (1975). *Report of the Committee on Mentally Abnormal Offenders. Cmnd. 6244*. HMSO, London.

Ingram, M. I. (1979). Paedophilia: the participating victim. *British Journal of Sexual Medicine*, January 22–5; February 24–6.

MacCulloch, M. J., Snowden, P. R., Wood, P. J. W., and Mills, H. E. (1983). Sadistic fantasy, sadistic behaviour and offending. *British Journal of Psychiatry*, **143**, 20–9.

Monahan, J. (1981) *Clinical prediction of violent behaviour*. Government Printing Office, Washington DC.

Monahan, J. (1989). Risk assessment of violence among the mentally disordered: generating useful knowledge. *International Journal of Law and Psychiatry*, **11**, 249–57.

Post, R. D., Willett, A. B., Franks, R. D., House, R. M., Black, S. M., and Weissberg, M. D. (1980). A preliminary report on the prevalence of domestic violence among psychiatric inpatients. *American Journal of Psychiatry*, **137**, 974–5.

Sapsford, R. J., Banks, C., and Smith, D. D. (1978). Arsonists in prison. *Medicine, Science and the Law*, **18**, 247–54.

Scott, P. D. (1977). Assessing dangerousness in criminals. *British Journal of Psychiatry*, **131**, 127–42.

Smith, S. M. and Hanson, R. (1974). 134 battered children: a medical and psychological study. *British Medical Journal*, **3**, 666–70.

Soothill, K. L. and Pope, P. J. (1973). Arson: a 20 year cohort study. *Medicine, Science and Law*, **13**, 127–38.

Soothill, K. L., Jack, A., and Gibbens, T. C. N. (1976). Rape: a 22 year cohort study. *Medicine, Science and the Law*, **16**, 62–9.

Soothill, K. L., and Gibbens, T. C. N. (1978). Recidivism of sexual offenders: a reappraisal. *British Journal of Criminology*, **18**, 267–76.

Taylor, P. J. and Gunn, J. (1984). Violence and psychosis. I: Risk of violence amongst psychotic men. *British Medical Journal*, **288**, 1945–9.

11

Chronic schizophrenia: can one do anything about persistent symptoms?

EVE C. JOHNSTONE

Introduction

The term 'chronic' is not always easy to define. In relation to disease in general it is used to convey various meanings: of long duration, severe, or unresponsive to treatment. In the context of schizophrenia it is also sometimes used to mean 'characterized by a defect state'. In this chapter it is being used to mean non-acute. There is likely to be some difference of opinion among clinicians about the point at which a symptom would be regarded as persistent. Examples of differing views of this kind may be found in a series of reviews on treatment resistance in schizophrenia (Dencker and Kulhanek 1988), where one group of authors use the term 'treatment resistance' to refer to the 6–8 per cent (Tuma and May 1979; Macmillan et al. 1986) of schizophrenic patients who appear to show no real response to treatment maintained over months and years; whereas another points out that 'treatment resistance' to social and vocational rehabilitation is the hallmark characteristic of a majority of schizophrenic patients including those designated as 'good responders' to in-patient treatment and therefore advocates that definitions of treatment resistance should be greatly extended. In this chapter, the term 'persistent symptoms' refers to symptoms which continue in spite of conventional psychiatric management.

It has become customary to classify the typical abnormalities of the mental state of schizophrenic patients into positive and negative symptoms with reference to behavioural excesses and deficits: positive symptoms are pathological by their presence and negative symptoms represent the loss of some normal function. Positive symptoms are generally considered to include delusions, hallucinations, incoherence of speech, and incongruity of affect. Although definitions of negative symptomatology concern deficits in behaviour, not all deficits are included and flattening of affect and poverty of speech are central to most definitions. Strictly speaking, incoherence of speech, incongruity of affect, flattening of affect, and poverty of speech are signs rather than symptoms, but they will be included in this account. In addition to positive and negative symptoms, schizophrenic patients may suffer from other symptoms such as depression or anxiety and, like the sufferers from other chronic disorders, may be very distressed about the implications of their condition. These non-specific symptoms will not however be discussed in this chapter, which concerns symptomatology characteristics of schizophrenia.

Positive symptoms

If current diagnostic criteria for schizophrenia are employed, all patients given this diagnosis must at some time suffer from positive symptoms. These symptoms are characteristic of acute episodes of schizophrenia and the efficacy of neuroleptic drugs in treating such episodes has been demonstrated many times, a particularly well-conducted study being that of the Collaborative Study Group of the National Institute of Mental Health (1964). Some patients, however, do not respond to neuroleptic drugs in this situation. For example, 17 of the 253 patients in the study of first episodes of schizophrenia conducted at Northwick Park Hospital did not achieve discharge within the two-year follow-up (Macmillan *et al*. 1986). This was in most instances due to the persistence of positive symptoms in the face of at least adequate neuroleptic treatment. To my knowledge two of those patients have had constant persistent positive symptoms now for more than six years although they have had treatment throughout this time. Of the patients in the Northwick Park functional psychosis study (Johnstone *et al*. 1988) who conformed to the DSM-III criteria for schizophrenia or schizophreniform psychosis and who were on active pimozide and completed the four weeks of the study ($n = 33$), there was a highly significant improvement in positive symptoms (mean week $0 = 8$: mean week $4 = 1.9$). However three patients still had one positive symptom with a maximal score and 11 had one positive symptom with a morbid score at the end of four weeks of treatment.

Symptoms which do not show any improvement in four weeks might not be classed by all as being persistent symptoms but most of us would be beginning to worry. What steps can reasonably be followed in this situation? These issues are dealt with in detail by the authors who have contributed to the book *Treatment resistance in schizophrenia* (Dencker and Kulhanek 1988) and in their recommendations it is stated that for many schizophrenic patients with persistent symptoms, treatment could be vastly improved by simply applying what we now know. This type of statement sounds optimistic but it does of course imply that many clinicians are not making proper application of current knowledge and I am far from sure that this is justified. None the less in this situation the issues that must be addressed are:
(1) the lines of action recommended in this situation;
(2) the basis for thinking that such action may be effective.
Recommended courses of action are considered in the sections that follow.

Concerning neuroleptic ingestion

The first step in this case is to ensure compliance. Patients do not always take the drugs that are prescribed even when in-patients. Where there is any question of doubt about this, liquid or parenteral medication should be substituted for tablets.

Secondly, the dosage of the drug must be considered. While it is true that across groups of patients very high doses appear to have no advantage over standard doses, it is clearly worth increasing the dose until there is at least some evidence (in terms of improvement or extra-pyramidal effects) that the drug is affecting the CNS.

All drugs which are known to be effective neuroleptics block dopamine receptors and elevate prolactin secretion. On this basis it could be thought that there is no reason, other than a dosage effect, why any individual should respond better to any one neuroleptic than to any other. None the less some patients do seem to respond better to one neuroleptic than to equivalent dosages of another. The basis for this was demonstrated in a study by McCreadie *et al.* (1984). In this study eleven male chronic schizophrenics were serially given four oral neuroleptics and two neuroleptics by depot intramuscular injection. There was considerable intra-individual variation in plasma levels of the different antipsychotic agents. These findings suggest that if a patient fails to respond to one neuroleptic even though adequate dosages have certainly been given, there may be good pharmacokinetic reasons for switching him to another belonging to a different group or for giving the same neuroleptic by a different route of administration.

At one time the idea was expressed that there could be a 'therapeutic window' regarding antipsychotic dosage and that not only would doses below a certain level be ineffective, but so too would levels above an optimum level. There has not been continued support for this view in the literature and indeed the relationships between neuroleptic blood levels and antipsychotic effect are in general too weak to be clinically useful. None the less in a patient on high doses of high potency neuroleptics a reduction in the dosage can be helpful. Neuroleptic-induced behavioural toxicity (Gelenberg and Mandel 1977) which may resemble severe, particularly catatonic schizophrenia has been described. Apart from this—as in general medicine—a patient who is not doing well and has over the weeks and months got into an increasingly elaborate drug regime (e.g., high doses of more than one neuroleptic, an anticholinergic, possibly a hypnotic and/ or antidepressant) can derive benefit from having as many of the drugs as possible stopped so that the regime is simplified and the dosages reduced. It is certainly possible for toxic psychotic symptoms to be superimposed upon schizophrenia. Neuroleptics being dopamine-blocking agents, all increase prolactin levels. Numerous studies have examined the relationship between prolactin levels and antipsychotic response. With large groups of patients, significant positive relationships are found but they are not sufficiently strong to be of value in the individual case. I have, however, used prolactin levels as evidence of non-compliance in a patient who has stated that he was taking oral drugs when he has not been doing so and this can be useful in supporting the case of compulsory treatment. All antipsychotic agents that are known to be effective block D_2 receptors and raise prolactin levels. Clozapine and sulpiride are sometimes thought of as being different. Clozapine has weak effects upon these receptors although it has been considered that its mode of action might differ from that of other neuroleptics [Kane *et al.* 1988). There is some evidence that it is of value in treating schizophrenic patients who are not doing well but its use is to some extent limited by the fact that it is associated with the development of agranulocytosis much more often than other standard neuroleptic drugs (Kane *et al.* 1988). Sulpiride, a substituted benzamide, blocks D_2 receptors and has a marked prolactin increasing effect.

Diagnostic review

Reasonable clinical competence and care in making the original diagnosis is assumed, but where a patient is doing badly he deserves a doctor who is prepared to question his/her own judgement and start all over again from scratch. In the 268 patients referred to Northwick Park Hospital for our first episodes of schizophrenia study there were 15 cases with organic illness of definite or possible aetiological significance for their psychosis (Johnstone *et al.* 1987*a*) and specific treatment would have been applicable for some of these disorders (see Table 9).

A similar percentage (24 of 328) was found in relation to our next study of functional psychoses (Johnstone *et al.* 1988) and again specific treatment was applicable in some cases (see Table 9). Careful diagnostic review is not going to reveal other treatable disorders in many cases but there are certainly enough of these for this step never to be overlooked.

Alternative forms of physical treatment

Lithium   Some studies have suggested that lithium is of value in the treatment of schizophrenia and schizo-affective disorders and it has also been reported that a combination of lithium and neuroleptic is better than neuroleptic alone in the treatment of schizophrenia. In a brief review concerning the issue of lithium treatment in schizophrenia and schizo-affective disorders, Prien (1979) referred to the view that schizophrenic patients who respond to lithium are in reality 'atypical' manic depressives and that a favourable response to lithium indicates a need for rediagnosis but pointed out that studies to date do not have adequate sample sizes or sufficient controls for a judgement on the merits of this or alternative positions. In a recent trial at Northwick Park Hospital (Johnstone *et al.*

Table 9   *Organic diagnosis in patients initially presenting with 'functional' psychotic pictures at Northwick Park Hospital*

Individual organic diagnosis	No. of patients from 268 cases referred as probable first schizophrenic episodes (Johnstone *et al.* 1987*a*)	No. of patients from 328 cases referred with psychotic illness (Johnstone *et al.* 1988)
Syphilis	3	1
Alcohol excess	3	2
Sarcoidosis	2	—
Drug abuse	2	11
Carcinoma of the lung	1	1
Autoimmune disease	1	1 (Systemic lupus erythematosus)
Epilepsy resulting from cerebral cysticercosis	1	—
Thyrotoxicosis	1	1
Head injury	1	
Hypothyroidism		1
Cerebro-vascular accident		1
Cerebral tumour		2
Ulcerative colitis on steroids		1
B_{12} deficiency		1
Uncontrolled insulin dependent diabetes mellitis		1

1988) we found that pimozide was equally effective in reducing psychotic symptoms in functionally psychotic patients with depressed or elevated mood or with no consistent mood change but the only significant effect of lithium was in reducing elevated mood. There was no evidence using PSE/Catego and DSM-III grouping that any subgroup (e.g., schizo-affective) according to specified criteria could be selected in which a significant benefit of lithium on positive psychotic symptoms could be found.

Antidepressants   While antidepressants are used in schizophrenia they are not normally given with a view to treating positive symptoms. Any benefits that they do provide tend to be in the direction of relief of superimposed depressive features rather than of schizophrenic symptomatology.

Other drugs   A number of other groups of drugs have been suggested for use in schizophrenia but there is either no real evidence of efficacy or the available evidence is conflicting.

Carbamazepine, a potent anticonvulsant, is widely used in the treatment of paroxysmal pain syndromes and has lately been increasingly used in the management of affective disorders (see pp. 33 and pp. 111). It has also been suggested as a treatment for schizophrenia, in particular in the management of patients with very disturbed behaviour (Post *et al.* 1986). The limited nature of the evidence for the efficacy of this drug in schizophrenia is presented in a review by Donaldson *et al.* (1983) in which other unconventional treatments of schizophrenia, including propranolol, clonidine, and benzodiazepines are also discussed and it is concluded that none of these has out-performed conventional neuroleptics in treating schizophrenia. Small studies have been described using endorphines and opiate blocking agents (Mueser and Dysken 1983) but the value of these drugs has not been demonstrated and the neuropeptide cholecystokinin (which has been found to coexist with dopamine in certain mesencephalic neurons) has not been shown to be effective in the treatment of schizophrenia (Tamminga *et al.* 1986).

Electroconvulsive therapy   Although it was as a treatment for schizophrenia that electroconvulsive therapy (ECT) was introduced in the 1930s, in recent decades it has been much more widely used in the treatment of depressive illness than in the management of schizophrenia, and there have been relatively few controlled trials of ECT in schizophrenia. The most recent were conducted by Taylor and Fleminger (1980) and Brandon *et al.* (1985) and both found that the addition of real ECT to neuroleptic medication had a greater effect than the addition of simulated ECT although in both cases the differences did not persist at 12-week follow-up. By contrast, the earlier trial of Miller *et al.* (1953) showed no difference between real and simulated ECT groups (see Chapter 12).

Non-physical treatments

Social interventions   These have been widely used in an attempt to affect the course of schizophrenia. Some studies have indicated that they may reduce the relapse rate in schizophrenia and others have suggested that their effect is more in terms of postponing relapse/re-admission. These techniques are not, however, generally used as a means of treating persistent positive symptoms.

Psychological techniques   Treatments using psychological principles have been directed at symptoms, but most claims for efficacy are based on single case studies and there is little evidence of their value from controlled trials (Bebbington and Kuipers 1987). Operant techniques have been used to

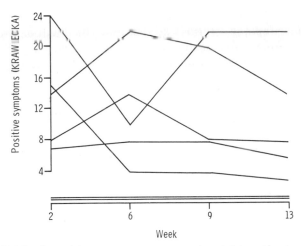

Fig. 4. Individual positive symptom scores in eight patients with chronic schizophrenia during 13 weeks of stable regime (Krawiecka scale ×2)

improve speech abnormalities and the amount of speech with delusional content, and it is claimed that it is sometimes possible to modify delusional beliefs using a non-confrontational approach and using thought-stopping techniques.

Concluding remarks on positive symptoms

The value of all these treatments of occasional use and which have been studied in terms of few or single cases and have not been subject to controlled clinical trials has to be considered against the fact that positive symptoms which by any criteria are persistent may show quite considerable changes. This was recently brought home to us in a study we carried out over eight months on eight of the most severely ill schizophrenic patients in Shenley Hospital. The mental states of the patients were assessed every two weeks for the eight months of the trial and in the 13-week run-in period where the treatment had been stable for months. These patients were never well but their positive symptoms showed substantial fluctuations (Fig. 4). It would be very easy for a single case study or uncontrolled trial to show apparently very promising results in this situation.

Negative symptoms

Although neuroleptics may be recommended as being likely to improve positive symptoms in acute episodes of most cases of schizophrenia, no treatment can be recommended as affecting negative symptoms in such a way. Negative symptoms do not of course occur in acute episodes in the way that positive symptoms do and indeed not all schizophrenic patients necessarily have them at all. It is said that fewer than half of schizophrenic patients show prominent negative features at any one time (Pogue-Geile and Zubin 1988) and in a study of 510 patients with schizophrenia and receiving long-term in-patient care for that condition, 180 showed non-morbid levels of negative symptomatology (Owens and Johnstone 1980). Positive symptoms have been said to be relatively variable and negative relatively fixed. While it is in general difficult to separate the question of reversibility/stability from responsivity to treatment in some sense, studies (Pfohl and Winokur 1982; Johnstone *et al.* 1987*b*) have shown that negative features are more stable than positive but that they are not invariably persistent. Few methods are available which can be reliably recommended to reduce negative symptoms.

Concerning neuroleptic ingestion

The effect of neuroleptics upon these symptoms is uncertain. At present, controlled trials have yielded evidence that neuroleptic medications may improve, have no significant effect on, or may exacerbate negative symptoms (Pogue-Geile and Zubin 1988). To some extent these inconsistencies may relate to the difficulties in rating negative symptoms. Because of the lack of convincing evidence that this group of drugs works at all in negative symptoms, adjustment of dosage and change of neuroleptics, as recommended under the section on positive symptoms, are probably not worthwhile as far as negative symptoms are concerned.

Diagnostic review

Negative symptoms may closely resemble the phenomenology of other psychiatric disorders, particularly depression and the akinesia induced by neuroleptic drugs. The occurrence of post-psychotic depression has long been noted. Lack of interest, slowness, and paucity of speech are well-established features of depression and rate on assessments of that disorder; but they would clearly also rate on assessments intended for negative schizophrenic features. The similarities between drug-induced akinesia and schizophrenic defect and depression have been well described (Rifkin *et al.* 1975). Where there is any real case for thinking that the symptoms may be depressive, adequate antidepressant treatment should be given and where the drug-induced akinesia is considered a serious possibility, the neuroleptics should be reduced or withdrawn.

Alternative forms of physical treatment

L-dopa, amphetamines, and anti-parkinsonian medications have all been tried as treatments for negative symptoms but the evidence that they are clearly effective is so far lacking (for review see Pogue-Geile and Zubin 1988). Certainly our own recent study has not shown any effect of L-dopa

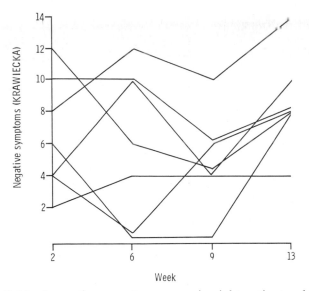

Fig. 5. Individual negative symptom scores in eight patients which chronic schizophrenia during 13 weeks of stable regime (Krawiecka scale × 2)

upon conventional ratings of negative symptoms. Some studies have shown apparent benefits for these and other treatments but the evidence is conflicting. Although negative symptoms have been shown to be more stable than positive even in very severe cases (Johnstone *et al.* 1987*b*), they may fluctuate as is shown in the 13-week run-in period of our recent study of very severely ill schizophrenic patients described above. The negative symptoms in the patients in this trial are shown in Fig. 5.

Non-physical treatments

Social treatments   Negative symptoms can be affected by social factors (Wing and Brown 1970), being worsened by an impoverished social environment and improved in some cases and to some extent by social stimulation. Once established, however, negative symptoms are likely to persist to some degree even in the most satisfactory social environment (Wing 1978).

Psychological treatments   The token economy is a specific application of operant methods. It is implemented particularly to modify impairments of social performance. Although it does seem to work in some cases, simple social reinforcement may be as effective. The approach has its limitations: some patients do not respond, some show deterioration in untargeted function and the benefits often fail to generalize (Bebbington and Kuipers 1987). Social skills training may be helpful in some schizophrenic patients but results again have been inconsistent.

Clinical guidelines

1. Consider whether the principal problem is persisting positive symptoms, negative symptoms or both.

2. Regarding positive symptoms:

    (i)   ensure compliance with neuroleptic treatment;

    (ii)   review the drug dosage and consider whether it has been high enough;

    (iii) consider a change to a neuroleptic of a different group or to the same neuroleptic given by a different route;

    (iv)   if the neuroleptic dosages are high, consider a reduction in dose;

    (v)   where the patient is on a complex drug regime (e.g., one or more neuroleptics and anticholinergics and antidepressants) stop as many of the drugs as possible;

    (vi)   diagnostic review with particular emphasis on the exclusion of organic disorders;

    (vii) consider alternative drug treatments remembering that none of these has out-performed conventional neuroleptics in treating positive symptoms of schizophrenia;

    (viii) consider the possibility that ECT may be of benefit remembering that there is no evidence of lasting benefit in schizophrenia;

    (ix)   consider the possibility of treatments using psychological principles in the case of the particular persistent symptoms suffered by this patient remembering that most claims for efficacy are based on single case studies.

3. Regarding negative symptoms:

    (i)   drug effects are much less certain than in positive symptoms—in the individual case there may be some value in first increasing the neuroleptic dose and if this has no benefit then reducing it below the initial level;

    (ii)   diagnostic review with particular emphasis on the possibility of depression or drug-induced akinesia;

    (iii) consider alternative drug treatments remembering that the evidence for their efficacy is so far lacking;

    (iv)   modify the patient's social environment to enhance the degree of social stimulation;

    (v)   consider the possibility of psychological treatments such as social reinforcement, social skills training, or token economy.

4. In the case of all persistent symptoms remember that in patients with

persistent symptoms of all kinds continued attempts to produce even minor improvements in terms of symptom reduction, improved function, or lessened side-effects are often much appreciated.

References

Bebbington, P. E. and Kuipers, L. (1987). Non-physical treatment of the psychoses. *British Medical Bulletin*, **43**, 704–17.

Brandon, S., Cowley, P., McDonald, C., Neville, P., Palmer, R., and Wellstood-Easen, S. (1985). Leicester ECT trial: results in schizophrenia. *British Journal of Psychiatry*, **146**, 177–83.

Dencker, S. J. and Kulhanek, F. (ed.) (1988). *Treatment resistance in schizophrenia*. S. J. Dencker and F. Kulhanek, Wiesbaden.

Donaldson, S. R., Gelenberg, A. G., and Baldessarini, R. J. (1983). The pharmacologic treatment of schizophrenia: a progress report. *Schizophrenia Bulletin*, **9**, 504–27.

Gelenberg, A. L. and Mandel, M. R. (1977). Catatonic reaction to high potency neuroleptic drugs. *Archives of General Psychiatry*, **34**, 947–50.

Johnstone, E. C., Macmillan, J. F., and Crow, T. J. (1987*a*). The occurrence of organic disease of possible or probable aetiological significance in a population of 286 cases of first episode schizophrenia. *Psychological Medicine*, **17**, 371–9.

Johnstone, E. C., Owens, D. G. C., Frith, C. D., and Crow, T. J. (1987*b*). The relative stability of positive and negative features in chronic schizophrenia. *British Journal of Psychiatry*, **150**, 60–4.

Johnstone, E. C., Crow, T. J., Frith, C. D., and Owens, D. G. C. (1988). The Northwick Park 'functional' psychosis study: diagnosis and treatment response. *Lancet*, **ii**, 119–25.

Kane, J., Honigfeld, G., Singer, J., Meltzer, H. Y., and the Clorazil Collaborative Study Group (1988). Clozapine for the treatment-resistant schizophrenic. *Archives of General Psychiatry*, **45**, 789–96.

McCreadie, R. G., Mackie, M., Wiles, D. H., Jorgensen, A., Hansen, V., and Menzies, C. (1984). Within individual variation in steady state plasma levels of different neuroleptics and prolactin. *British Journal of Psychiatry*, **144**, 625–9.

Macmillan, J. F., Crow, T. J., Johnson, A. L., and Johnstone, E. C. (1986). The Northwick Park study of first episodes of schizophrenia. III Short term outcome in trial entrants and trial eligible patients. *British Journal of Psychiatry*, **148**, 128–33.

Miller, D. H., Clancy, J., and Cumming, E. (1953). A comparison between uni-directional current non-convulsive electrical stimulation given with Reiter's machine standard alternating current electroshock (Cerletti method) and pentolal in chronic schizophrenia. *American Journal of Psychiatry*, **109**, 617–21.

Mueser, K. T. and Dysken, M. W. (1983). Narcotic antagonists in schizophrenia: a methodological review. *Schizophrenia Bulletin*, **9**, 213–25.

National Institute of Mental Health, Psychopharmacology Service Centre, Collaborative Study Group (1964). Phenothiazine treatment of acute schizophrenia. *Archives of General Psychiatry*, **10**, 246–61.

Owens, D. G. C. and Johnstone, E. C. (1980). The disabilities of chronic schizophrenia: their nature and the factors contributing to their development. *British Journal of Psychiatry,* **136,** 384–95.

Pfohl, B. and Winokur, G. (1982). Schizophrenia: course and outcome. In *Schizophrenia as a brain disease* (ed. F. A. Henn and H. A. Nasrallah), pp. 26–39. Oxford University Press.

Pogue-Geile, M. F. and Zubin, J. (1988). Negative symptomatology and schizophrenia: a conceptual and empirical review. *International Journal of Mental Health,* **16,** 3–45.

Post, R. M., Rubinow, D. R., Uhde, T. W., Ballenger, J. C., and Linnoila, M. (1986). Dopaminergic effects of carbamazepine. *Archives of General Psychiatry,* **43,** 392–6.

Prien, R. J. (1979). Lithium in the treatment of schizophrenia and schizo-affective disorders. *British Journal of Psychiatry,* **141,** 387–400.

Rifkin, A., Quitkin, F., and Klein, D. F. (1975). Akinesia. *Archives of General Psychiatry,* **32,** 672–4.

Tamminga, C. A., Littman, R. L., and Alphs, L. D. (1986). Cholecystokinin: A neuropeptide in the treatment of schizophrenia. *Psychopharmacology Bulletin,* **22,** 129–31.

Taylor, P. and Fleminger, J. J. (1980). ECT for schizophrenia. *Lancet,* **i,** 1380–3.

Tuma, A. H. and May, P. R. A. (1979). And if that doesn't work, what next . . .? A study of treatment failures in schizophrenia. *Journal of Nervous and Mental Disease,* **167,** 566–71.

Wing, J. K. (1978). Clinical concepts of schizophrenia. In *Schizophrenia: towards a new synthesis* (ed. J. K. Wing), pp. 1–30. Academic Press, London.

Wing, J. K. and Brown, G. W. (1970). *Institutionalism and schizophrenia.* Cambridge University Press, London.

12

Schizophrenia and ECT: a case for change in prescription?

PAMELA J. TAYLOR

Historical background

Electroconvulsive therapy (ECT) is just over 50 years old. The basis of its introduction, namely the theory that fits and schizophrenia might be biologically incompatible, has suffered some of the same vicissitudes as the treatment it spawned. The hypothesis was exciting and the results initially promising. Currier *et al.* (1952) found that not only the individual prognosis of schizophrenia dramatically improved, but that the general attitude to people with the illness was transformed as a result. ECT was not, however, consistently helpful and therapeutic excitement was transferred to the new drug treatments. When Slater *et al.* (1963) showed that schizophreniform illnesses and epilepsy were not incompatible, and, in fact, that there was a greater than chance expectancy of their coexistence, perhaps disillusion with ECT for schizophrenia should have been complete. Reynolds (1968), however, noted that in patients with both conditions an increase in fit frequency tended to occur as the psychosis improved, and conversely that florid psychosis developed with seizures under good control. Landolt (1958) had coined the term 'forced normalization' to describe the latter phenomenon. By the 1970s it was clear that neuroleptic drugs were not consistently curative and could carry important short and long-term risks (Gardos and Cole 1976). A growing belief among practitioners that there might be a case for re-evaluating a role for ECT in schizophrenia was, if anything, reinforced by the stridency of the emotive and poorly informed views of the antipsychiatrists of the period (*British Medical Journal* 1975).

Patterns of clinical choice

The use of ECT has been generally declining throughout the developed world over the last 20 years, although there is some evidence that in the USA its use is on the increase again, particularly for the elderly or physically ill in whom antidepressant medication may be dangerous (Abrams 1988). There seems to have been a disproportionate decline in its use for schizophrenia. A Finnish study (Miskanen and Achte 1972) found that by 1950, 55 per cent of an admission sample of people with schizophrenia received

143

ECT, compared with 14 per cent just 15 years later. A Canadian study, covering a slightly later period, up to 1973, found that 15 per cent of all patients receiving ECT had a diagnosis of schizophrenia (Eastwood and Peacocke 1976).

The American Psychiatric Association (APA) survey   Responders to the APA Task Force Survey (1978, pp. 1–12) indicated that only major depressive illness ranked above schizophrenia as a reason for giving ECT, but while 77 per cent of patients actually receiving it in the previous six months had a major depression, only 17 per cent had schizophrenia.

Discrepancy between public and private prescription standards: cost effectiveness   ECT was outlawed altogether in some state hospitals, such as those of Alabama, but only because of the poor conditions at the time in those hospitals (American Psychiatric Association 1978, pp. 137–139). By contrast, 56 per cent of a hundred consecutive first admissions for schizophrenia to one private facility were treated with ECT (Sullivan 1974). Should such discrepancy be interpreted with cynicism? Perhaps patients in private care were getting ECT simply as the cheapest available option. If this were really so, however, then the high use of ECT in the private sector might represent a hard and practical, if indirect, validation of more scientifically established findings. Early researchers reported reduction in recovery time and length of hospitalization for schizophrenia with ECT.

In developing countries, the use of ECT for schizophrenia has remained higher for longer, also largely on grounds of cost effectiveness (Doongaji *et al.* 1973).

The British survey   In Britain, where prescription of ECT is neither blocked completely by laws, nor yet generally dictated by fiscal considerations, data on prescription by diagnosis are not available, but an opinion survey is instructive (Pippard and Ellam 1981). In 1979, just 17 psychiatrists, of 2198 responders nationally, said that ECT would often be appropriate for chronic schizophrenia. For all the other functional psychotic illnesses, including acute schizophrenia, ECT was considered to be at least occasionally appropriate. The figures also show, however, that although trainees would follow consultant patterns of using ECT for schizo-affective psychosis and the affective psychoses, they appeared to dissent on its use for acute schizophrenia, probably avoiding it altogether. Were trainees more susceptible to pressure groups and the threat of legalistic struggles, given that people with acute schizophrenia would form the patient group least likely to be competent to make the treatment decision? Or were they at the front line of evidence on efficacy?

The clinical efficacy of ECT for schizophrenia

Early studies: the first two decades

Most early literature suggested that ECT was beneficial for schizophrenia, but few reports can now be taken as wholly satisfactory. The problems in their interpretation include the following.

1. *Their open or anecdotal nature*   However dramatic, anecdotes are *not* enough; it is almost axiomatic that published treatment outcome anecdotes will be favourable. Uncontrolled studies may not avoid observer bias. ECT is peculiarly subject to proselytizers and antagonists alike, and thus requires exceptional adherence to scientific method.

2. *Lack of diagnostic standardization*   Many studies were from America at a time when there were considerable international discrepancies in diagnostic practice (Kendell 1971). Of specific interest was the broader concept of schizophrenia in the USA, embracing much that would be regarded as affective disorder in Britain. As ECT may be the treatment of choice for some affective psychoses, the difficulties in interpreting the schizophrenia literature of this period became considerable.

3. *The changing nature of schizophrenia*   Catatonic schizophrenia seemed particularly susceptible to ECT. Currier *et al.* (1952) reported that nearly 30 per cent of their series of patients receiving ECT with success in 1946 had catatonic schizophrenia, and such a proportion was not unusual. Little catatonic schizophrenia is now seen in the western world (Morrison 1974). Studies which did or could have included a substantial proportion of people with catatonia may be misleading for current practice. There is the possibility, too, that diagnostic difficulties were particularly great in distinguishing between the affective and schizophrenic variants of excitement and stupor.

4. *Underdeveloped methods for rating change*   Discharge from hospital, or re-admission to hospital, were not uncommonly indications of change that were used in isolation from descriptions of pure clinical change. Schwartz *et al.* (1975) showed that hospitalization tended to correlate well with such variables as social class or family position, but to bear little relation to clinical state.

The early double-blind placebo-controlled trials

None of these trials showed ECT offering an advantage for chronic schizophrenia (Miller *et al.* 1953; Brill *et al.* 1959, Heath *et al.* 1964), but the one study which included an unspecified number of people with acute schizophrenia did show significantly favourable results for the convulsive therapies (Ulett *et al.* 1956).

Clinical responsiveness: avoiding ECT at acute and chronic extremes

May (1968) deliberately selected a group of patients with a middle range of disorder, that is, neither suffering from chronic schizophrenia nor the very acute, reactive, and rapidly recovering form. ECT was supplemented only by 'milieu therapy' when compared with other treatments. There was no attempt at double-blind evaluation because of the very obvious differences in the treatments. ECT alone brought improvement in about 80 per cent of the treatment group, thus occupying an intermediate position between the drug therapies and the more psychological approaches, the latter including the dynamic psychotherapies. May and colleagues (1976a and b) then reported the progress of these patients for up to five years. It is difficult to know how to interpret these results, given the likely powerful and uncontrolled effects of so many treatments and events since the onset of the trial, but they found less and less to choose between the outcome for the drug and ECT groups over time, with a trend towards an advantage for the ECT group.

ECT-drug treatment combinations

Berg *et al.* (1959) established the safety of an ECT–phenothiazine combination without finding it advantageous. A number of other uncontrolled studies consistently demonstrated that ECT had a facilitating effect. The first controlled, but not double-blind study among newly admitted women with schizophrenia found a significant advantage for the ECT–drug combination over either ECT or drugs given alone (Childers 1964). Smith *et al.* (1967), working with people who were acutely ill, but with a four-year history, showed a significantly faster improvement rate in the ECT–drug combination group than in the drugs alone group. Only 12 months after the start of treatment had the latter group caught up in terms either of the percentage still hospitalized or global psychopathology ratings.

Janakiramaiah and colleagues (1982) reported on a group of patients recently admitted to hospital whose mean duration of illness was only 18 months, but who had had schizophrenic symptoms on this occasion for an average of 4.6 months. They conducted a single-blind study of chlorpromazine (CPZ) at two different dose levels (300 mg and 500 mg) compared with CPZ and ECT in combination. The major difference between their sample and the previous two studies was that all previous treatment had lapsed for some considerable, but unspecified, time before entry into the study. Thus, drugs and ECT were both new treatments for the episode. There was a complicated relationship between ECT and progress. The most important finding was that by the sixth week of treatment all patients had improved significantly from their admission state and there was no significant difference between the groups. In between, the advantage passed from group to group, almost at random. It might have been interesting to extend the follow-up period further, as there was a hint that the patients receiving 500 mg of chlorpromazine alone might have begun to show some deterioration.

The modern double-blind controlled trials

Taking the studies overall, if there were a real advantage in adding ECT to neuroleptic drugs, it seemed likely to be in an earlier improvement in symptoms. The evidence for a facilitating effect of ECT for people with schizophrenia receiving medication was mixed, but perhaps only those already responding poorly to neuroleptics experienced a facilitating effect. Witton (1962) was more specific about a striking improvement in disordered thought processes, for example, following ECT after a poor drug response. Given the remaining uncertainty a double-blind controlled trial seemed necessary, but had to break again the awkward barrier of deliberately withholding a treatment—ECT—for some patients which many practitioners believed to be useful and most previous researchers had found of value. Furthermore, it meant offering to other patients an appropriate placebo. A barbiturate at anaesthetic level seemed the only choice, thus presenting a placebo treatment believed to be without specific effect but with a definite, if small, risk.

The Taylor and Fleminger trial (1980) was the first. Despite the fact that only persistently psychotic people were approached, roughly equivalent to May's middle prognosis group, the vast majority demonstrated a remarkable grasp of what was being asked of them. Perhaps because of their illness, many were disconcertingly blunt and challenging in their expression of their understanding of the study that was carried out: e.g., patient: 'You mean you want me to be a guinea pig?'; researcher: 'Yes'; patient: 'All right then'.

In order to be sure that the diagnosis was correct and that rapid treatment responders were excluded, the assessment period was extended over several weeks, with a minimum of two weeks on a stable dose of neuroleptics. The majority had been stabilized for a longer period. The attrition rate was high. Of the 70 people initially assessed, 15 were not confirmed as having schizophrenia by the research criteria; 11 had to be excluded because of recovery from their illness; refusal to participate in the trial or absence of informed consent accounted for 16; and a variety of other reasons including absconding and clinical staff unease for the remaining eight. Of those refusing participation, only three refused to enter the trial, the remainder rejected in whole or in part their need for treatment of any kind. Patients entering the trial proper were stratified for gender and for depression scores and then randomly assigned to the two groups. Assessments were at regular intervals for at least four months. All the researcher assessments were blind, but supplemented by a few non-blind assessments from the clinical team.

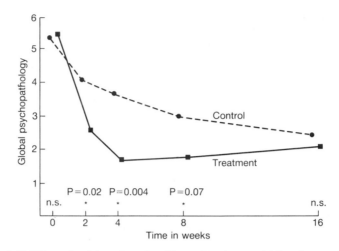

Fig. 6 ECT and schizophrenia. From Taylor and Fleminger 1980.

Overall, both groups improved significantly. The real ECT–drug combination group improved very significantly faster than the placebo ECT–drug combination grup (see Fig. 6). The equivalence of the improvement in the groups by the fourth month after the start of ECT was mostly due to 'catching up' by the drugs-only group. The plateauing of the ECT group was mainly accounted for by the marked deterioration of one patient. A check of symptomatic improvement showed a non-significant speeding of the recovery of depressive symptoms. The significant symptomatic improvements were in the more classically schizophrenic symptoms—namely, delusions of control, delusions of persecution, delusional mood, and thought interference.

Brandon and colleagues (1985) studied the problem completely independently of Taylor and Fleminger, but using very similar methodology and rating scales. Slightly fewer patients completed the trial—just nine completed the course of real ECT and eight the course of simulated ECT, both in combination with active medication. In order to compare the findings, Fig. 7 shows the changes in global psychopathology.

The real ECT patients were advantaged for up to four weeks after the start of treatment, which generally meant during the actual course of ECT. On the scale shown in Fig. 7 the differences between the groups just failed to reach significance but, when the items reflecting more typical schizophrenic symptoms were selected out using the (Montgomery *et al.* 1978) schizophrenia sub-scale from the Comprehensive Psychiatric Rating Scale (CPRS) (Asberg *et al.* 1978), the differences were significant. The relative improvement on the Hamilton Depression Scale (1960), by contrast, did not reach significance. Again, then, contrary to expectation, the more specifically schizophrenic symptoms improved most. Again, however, the

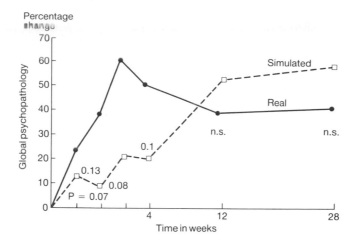

Fig. 7. ECT and schizophrenia. From Brandon *et al.* 1985.

group not receiving the real ECT caught up with the other group over time. Brandon noted that members of the group receiving real ECT generally needed less treatments.

The Abraham and Kulhara Study (1987) was performed at Chandigarh in India, and again was completely independent from the other studies, although the second author had been a clinical colleague of Taylor and Fleminger at the time of their trial. The design was similar, although in this case the freeze on treatment other than ECT lasted for the 16 weeks of the trial (instead of four), and the pre-trial stabilization period is not quite clear. Eleven patients in each group completed the trial. The results, as expressed by the Clinical Global Impression Scale, are shown in Fig. 8. The rating system was comparable but not exactly the same as that of the two previous studies. In this study, those patients receiving real ECT in addition to medication fared significantly better until the twelfth week, when the group receiving simulated ECT improved sufficiently to abolish statistically significant differences. The authors commented further that, as their patients did not score highly on depression scales at any point in the trial, it seemed most likely that the major effect of the ECT was on psychotic symptoms, rather than on mood.

The impact of the modern double-blind studies taken together
Each study was small. Taylor and Fleminger's, with an attrition rate of sample from 70 to 20, shows most clearly the difficulty of getting samples even of this size. As the methodology of each study was very similar and the findings entirely consistent, it might be said that, in effect, 30 patients treated with the real ECT as an adjunct to quite small doses of neuroleptic medication have been compared with 29 patients treated with placebo

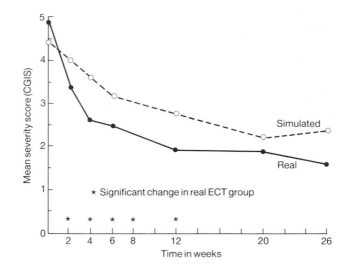

Fig. 8. ECT and schizophrenia. From Abraham and Kulhara 1987.

ECT, in conjunction with similar levels of medication. In these circumstances ECT undoubtedly accelerates improvement and the main effect of the treatment combination is on the most characteristic symptoms of schizophrenia. Patients receiving medication without ECT, however, do reach equivalent levels of recovery—at four weeks in the Leicester study, at eight weeks in the London study, and at sixteen weeks in the Chandigarh study. The patients were not, however, without adjunct to their medication. Could the barbiturate anaesthetic amount, after all, to more than placebo?

The report of Dodwell and Goldberg (1989) might be taken as an important addendum to the double-blind work. They studied 17 patients with a research diagnosis (RDC) of schizophrenia, or schizo-affective psychosis, before and after ECT, to test the strength of possible predictors of response culled from previous literature. Some, such as short duration of illness or episode, and absence of premorbid schizoid or paranoid personality traits, were associated with good outcome after ECT, although the duration link cannot really be settled outside double-blind conditions. Others, like depressive features, were found to be helpful predictors of outcome overall, but not specifically in relation to ECT. Patients expressed favourable attitudes to ECT.

ECT for schizophrenia: safety

Mortality from ECT

Mortality was the earliest safety consideration. No treatment known to medicine is without its mortality figures. Always low for ECT, these are now similar to those for other procedures requiring brief anaesthesia, such as tooth extraction (Heshe and Roeder 1976; Tomlin 1974). Mortality from schizophrenia is about 10 000 times more likely than mortality from an ECT.

The possibility of brain damage

This has received more recent attention. People with schizophrenia could, in theory, be at most risk because, in the 1950s and 1960s they tended to receive a higher number of treatments than those in other diagnostic groups. Two groups of workers independently examined the effects of large numbers of treatments for patients with chronic schizophrenia, in most cases the patients having received well over 100 treatments, to a maximum of 263 (Templer *et al.* 1973; Goldman *et al.* 1972). Allowing for severity of psychosis, some defects on psychometric testing were found, although caution has to be exercised in inferring too much from abnormalities in a single class of test, which was the principal finding. CT scan findings are equivocal. A British study (Johnstone *et al.* 1976) found no evidence of abnormality linked with ECT, but an American study (Weinberger *et al.* 1979) found significantly greater ventricular enlargement among those patients with schizophrenia who had had ECT, compared with those who had not. It is worth adding, however, that the characteristics of the schizophrenic illness also differed between the groups with different CT-scan findings. A more direct attempt to examine the possibility of brain damage secondary to ECT has focused on the measurement of creatinine phosphokinase (CPK) after ECT. CPK is an enzyme commonly released after damage to cells. No brain or heart CPK was released but small amounts of skeletal muscle CPK, compatible with the effects of muscle relaxant given before treatment, could be detected (Taylor *et al.* 1981).

Little of what is known about the impact of ECT on memory refers specifically to people with schizophrenia, but, aside from a possibility of larger numbers of treatments for schizophrenia in the early days of ECT, there is no reason to expect a memory effect would vary with diagnosis. There has been a great deal of recent evaluation of memory after ECT. Squire (1986) offers an excellent review of the field, including his own extensive work. In essence, no permanent effects on memory function have been definitely isolated, although Squire and his colleagues (1981) did find that some specific memories (for public events) apparently formed for up to three years before treatment remained impaired after ECT.

ECT for schizophrenia: the mechanism of action

There is insufficient space to consider fully ideas and evidence about the mechanism of action of ECT in schizophrenia. Although the original ideas of the incompatibility between fits and psychosis have been developed, modified, but never wholly discounted, the main interest now is in the biochemical effects. Evidence of prolactin release after ECT (e.g., Ohman *et al.* 1976; O'Dea *et al.* 1978) has been interpreted as part of a physiological stress response, rather than evidence of dopamine blockade. Animal work suggested that the efficacy of ECT depended on enhancement of dopamine and other amine responses (e.g., Grahame-Smith *et al.* 1978), but this seemed incompatible with the observations that, if not always restoring mental state to normal, ECT had never been implicated in an exacerbation of schizophrenic symptoms. Most recently, Cooper *et al.* (1988) have completed an open study which showed improvements in the positive symptoms of schizophrenia in twelve patients, accompanied by a time-dependent increase in the cerebro-spinal fluid concentration of homovanillic acid, a metabolite of the neurotransmitter dopamine. There were no changes in the metabolites of other brain amines. They argued that this may reflect a similar response to that of neuroleptic drugs, namely an initial blockade of dopamine receptors, after which a marked increase in dopamine turnover is seen, followed by tolerance.

Clinical guidelines

1. ECT should probably not be used in chronic schizophrenia. There is no evidence of specific therapeutic efficacy.
2. ECT should probably not be used in most very acute forms of schizophrenia. Many patients will respond rapidly to a change in their environment, or fairly small doses of medication.
3. ECT should be considered for patients showing:
   (a) acute or sub-acute catatonic features; or
   (b) moderate responsiveness to treatment, if
       (i) speed of recovery is particularly important, or
       (ii) the classically positive symptoms of schizophrenia are prominent and responding only slowly or partially to medication.
4. It is likely that the particular responsiveness to ECT of schizophrenia with a prominent affective loading is a myth that should be laid to rest.
5. There has been minimal research into the claim that aggression arising in the context of schizophrenia may be helped by ECT and any such use should be regarded as wholly experimental.
6. The use of maintenance ECT might bear further exploration. Improvement after treatment with neuroleptics is generally followed by maintenance drug therapy; improvement after ECT is almost never followed by maintenance ECT. Fink (1979, p. 207) considers that maintenance ECT poses no special risk, but there has been no systematic enquiry into its usefulness for schizophrenia.

7. Patients must, as far as possible, be allowed to make free choices about their treatment. This should include adequate information about the range and availability of treatments as well as a more detailed account of individual treatment a doctor proposes to recommend.

For schizophrenia the most balanced approach would probably be to tell all patients not responding to milieu adjustment, drugs, or psychological approaches within a few weeks or months that ECT might speed recovery, or even improve their chances of it. Tiny risks of mortality and morbidity are likely to be involved, which are certainly substantially less than for the untreated illness and probably substantially less than the risks of medication. Memory may be affected by treatment. Some patients have reported islands of memory loss which persist, but this is probably unusual.

Acknowledgements

My thanks to Joanne Crawford for preparing the manuscript.

References

Abraham, K. R. and Kulhara, P. (1987). The efficacy of electroconvulsive therapy in the treatment of schizophrenia. *British Journal of Psychiatry*, **151**, 152–5.

Abrams, R. (1988). *Electroconvulsive therapy*. Oxford University Press, New York.

American Psychiatric Association (1978). *Electroconvulsive therapy*. Task Force Report 14. APA, Washington DC.

Asberg, M., Montgomery, S. A., Perris, C., Schalling, D., and Sedvall, G. (1978). The Comprehensive Psychiatric Rating Scale. *Acta Psychiatrica Scandinavica*, Supplement **271**, 5–27.

Berg, S., Gabriel, A. R., and Impastato, D. J. (1959). Comparative evaluation of the safety of chlorpromazine and reserpine used in conjunction with ECT. *Journal of Neuropsychiatry*, **1**, 104–7.

Brandon, S., Cowley, P., McDonald, C., Neville, P., Palmer, R., and Wellstood-Eason, S. (1985). Leicester ECT trial: Results in schizophrenia. *British Journal of Psychiatry*, **146**, 177–83.

Brill, N. Q., Crumpton, E., Eiduson, S., Grayson, H. M., Hellman, L. I., and Richards, R. A. (1959). Relative effectiveness of various components of electro-convulsive therapy. *Archives of Neurology and Psychiatry*, **81**, 627–35.

British Medical Journal (1975). Anti-psychiatrists and ECT. *British Medical Journal*, **4**, 1–2.

Childers, R. T. (1964). Comparison of four regimes in newly admitted female schizophrenics. *American Journal of Psychiatry*, **120**, 1010–11.

Cooper, S. J., Leahey, W., Green, D. F., and King, D. J. (1988). The effect of electroconvulsive therapy on CSF amine metabolites in schizophrenic patients. *British Journal of Psychiatry*, **152**, 59–63.

Currier, G. E., Cullinan, C., and Rothschild, D. (1952). Results of treatment of schizophrenia in a state hospital. *Archives of Neurology and Psychiatry*, **67**, 80–8.

Dodwell, D. and Goldberg, D. (1989). A study of factors associated with response to electroconvulsive therapy in patients with schizophrenic symptoms. *British Journal of Psychiatry,* **154,** 635–9.

Doongaji, D. R., Jeste, D. V., Saoji, N. J., Kane, P. V., and Ravindranath, S. (1973). Unilateral versus bilateral ECT in schizophrenia. *British Journal of Psychiatry,* **123,** 73–9.

Eastwood, M. R. and Peacocke, J. E. (1976). Diagnosis and evaluation of ECT. *Canadian Psychiatric Association Journal,* **21,** 55–9.

Fink, M. (1979). *Convulsive therapy: theory and practice.* Raven Press, New York.

Gardos, G. and Cole, J. O. (1976). Maintenance antipsychotic therapy: is the cure worse than the disease? *American Journal of Psychiatry,* **133,** 32–6.

Goldman, H., Gomer, F. E., and Templer, D. L. (1972). Long-term effects of electroconvulsive therapy upon memory and perceptual-motor performance. *Journal of Clinical Psychology,* **28,** 32–4.

Grahame-Smith, D. G., Green, A. R., and Costain, D. W. (1978). Mechanism of the antidepressant action of electroconvulsive therapy. *Lancet,* **1,** 254–6.

Hamilton, M. (1960). A rating scale for depression. *Journal of Neurology, Neurosurgery and Psychiatry,* **23,** 56–62.

Heath, E. S., Adams, A., and Wakeling, P. L. G. (1964). Short courses of ECT and simulated ECT in chronic schizophrenia. *British Journal of Psychiatry,* **110,** 800–7.

Heshe, J. and Roeder, E. (1976). Electroconvulsive therapy in Denmark. *British Journal of Psychiatry,* **128,** 241–5.

Janakiramaiah, N., Channabasavanna, S. M., and Murthy, N. S. (1982). ECT/chlorpromazine combination versus chlorpromazine alone in acutely schizophrenic patients. *Acta Psychiatrica Scandinavica,* **66,** 464–70.

Johnstone, E. C., Crow, T. J., Frith, C. D., Husband, J. and Kreel, L. (1976). Cerebral ventricular size and cognitive impairment in chronic schizophrenia. *Lancet,* **2,** 924–6.

Kendell, R. E. (1971). Psychiatric diagnosis in Britain and the United States. *British Journal of Hospital Medicine,* **6,** 147–55.

Landolt, H. (1958). Serial electroencephalographic investigations during psychotic episodes in epileptic patients and during schizophrenic attacks. In *Lectures on epilepsy* (ed. A. M. Lorentz de Haas), pp. 91–133. Elsevier, Amsterdam.

May, P. R. (1968). *Treatment of schizophrenia.* Science House, New York.

May, P. R. A., Tuma, A. H., and Dixon, W. J. (1976*a*). Schizophrenia—a follow up study of results of treatment. I. Design and other problems. *Archives of General Psychiatry,* **33,** 474–8.

May, P. R. A., Tuma, A. H., Yale, C., Potepan, P., and Dixon, W. J. (1976*b*). Schizophrenia—a follow up study of results of treatment. II. Hospital stay over two to five years. *Archives of General Psychiatry,* **33,** 481–6.

Miller, D. H., Clancy, J., and Cumming, E. (1953). A comparison between unidirectional current, nonconvulsive electrical stimulation given with Reiter's machine, standard alternating current electroshock and pentothal in chronic schizophrenia. *American Journal of Psychiatry,* **109,** 617–20.

Miskanen, P. and Achte, K. A. (1972). The course and prognosis of schizophrenic psychoses in Helsinki: a comparative study of first admissions 1950, 1960, 1965. *Monographs from the Psychiatric Clinic of Helsinki University Central Hospital No. 4.,* Helsinki University.

Montgomery, S., Taylor, P. J., and Montgomery, D. (1978). Development of a schizophrenia scale sensitive to change. *Neuropharmacology*, **17**, 1061 3.

Morrison, J. R. (1974), Changes in subtype diagnosis of schizophrenia 1920–66. *American Journal of Psychiatry*, **131**, 674–7.

O'Dea, J. P. K., Gould, D., Halberg, M., and Wieland, R. G. (1978). Prolactin changes during electroconvulsive therapy. *American Journal of Psychiatry*, **135**, 609–11.

Ohman, R., Walinder, J. and Balldin, J. (1976). Prolactin response to electro-convulsive therapy. *Lancet*, **2**, 936–7.

Pippard, J. and Ellam, L. (1981). *Electroconvulsive treatment in Great Britain, 1980. A Report to the Royal College of Psychiatrists*. Gaskell, London.

Reynolds, E. H. (1968). Epilepsy and schizophrenia: relationship and biochemistry. *Lancet*, **1**, 398–401.

Schwartz, C. C., Myers, J. K., and Astrachan, B. M. (1975). Concordance of multiple assessments of the outcome of schizophrenia. *Archives of General Psychiatry*, **32**, 1221–7.

Slater, E., Beard, A. W., and Glithero, E. (1963). The schizophrenia-like psychoses of epilepsy. *British Journal of Psychiatry*, **109**, 95–150.

Smith, K., Surphlis, W. R. P., Gynthe, M. D., and Shimkunas, A. M. (1967). ECT–chlorpromazine and chlorpromazine compared in the treatment of schizophrenia. *Journal of Nervous and Mental Disease*, **144**, 284–92.

Squire, L. R., Slater, P. C., and Miller, P. C. (1981). Retrograde amnesia and bilateral electroconvulsive therapy. *Archives of General Psychiatry*, **38**, 89–95.

Squire, L. R. (1986). Memory functions as affected by electroconvulsive therapy. *Annals of the New York Academy of Sciences*, **462**, 307–14.

Sullivan, P. R. (1974). Treatment of acute schizophrenia: the place of ECT. *Diseases of the Nervous System*, **35**, 467–9.

Taylor, P. J. and Fleminger, J. J. (1980). ECT for schizophrenia. *Lancet*, **1**, 1380–2.

Taylor, P. J., Van Witt, R. J., and Fry, A. H. (1981). Serum creatine phosphokinase activity in psychiatric patients receiving electroconvulsive therapy. *Journal of Clinical Psychiatry*, **42**, 103–5.

Templer, D. I., Ruff, C. F., and Armstrong, G. (1973). Cognitive functioning and degree of psychosis in schizophrenics given many electroconvulsive treatments. *British Journal of Psychiatry*, **123**, 441–3.

Tomlin, P. J. (1974). Death in an out-patient dental anaesthetic practice. *Anaesthesia*, **29**, 551–70.

Ulett, G. A., Smith, K., and Gleser, G. C. (1956). Evaluation of convulsive and subconvulsive shock therapies utilizing a control group. *American Journal of Psychiatry*, **112**, 795–802.

Weinberger, D. R., Torrey, E. F., Neophytides, A. N., and Wyatt, R. J. (1979). Lateral cerebral ventricular enlargement in chronic schizophrenia. *Archives of General Psychiatry*, **36**, 755–9.

Witton, K. (1962). Efficacy of ECT following prolonged use of psychotropic drugs. *American Journal of Psychiatry*, **119**, 79–80.

13

Tardive dyskinesia: can it be prevented?

THOMAS R. E. BARNES

The tardive dyskinesia syndrome

Clinical features

Tardive dyskinesia is recognized as a movement disorder associated with the administration of dopamine antagonists, most commonly antipsychotic drugs. The syndrome was first described by Schonecker (1957) within five years of the introduction of antipsychotic drugs. The core feature is oro-facial dyskinesia, being the most familiar and prevalent sign of the condition. The early descriptions emphasized involuntary movements of the tongue, jaw, and face, such as protrusion or twisting of the tongue, lip smacking, puffing of the cheeks, pursing and sucking actions of the lips, chewing or lateral jaw movements, and movements of the peri-orbital muscles. In addition to oro-facial movements, abnormal involuntary movements of the trunk and limbs, usually described as choreiform or choreoathetoid, are commonly included within the syndrome.

Patients rarely complain spontaneously of tardive dyskinesia, being apparently unaware of the movements in most cases. In mild cases, friends and relatives of the patient are often not conscious of the movements as a drug side-effect or a manifestation of psychiatric illness, accepting them as part of the person's behaviour. This may reflect that such movements are familiar as a natural consequence of ageing in that a syndrome of idiopathic or spontaneous oro-facial dyskinesia, indistinguishable from tardive dyskinesia is seen in 5–15 per cent of elderly people who have never received antipsychotic drugs (Kane et al. 1982; Blowers et al. 1982).

Prevalence, severity, and natural history

The syndrome is usually considered to be a chronic complication of long-term treatment with antipsychotic medication. Comprehensive reviews of the relevant literature have concluded that there has been a true increase in the prevalence of tardive dyskinesia in the last 20 years. However, the reported prevalence figures vary widely, from 0.5 to 56 per cent, with a mean of around 20 per cent (Kane and Smith 1982), although figures of over 50 per cent are common in elderly populations.

157

In a major prospective study of tardive dyskinesia, Kane *et al.* (1985) found an incidence of 4 per cent a year over the first six years of treatment with antipsychotic drugs. Although patients may be at a relatively high risk of developing the condition in the first few years of drug therapy (Toenniessen *et al.* 1985), nevertheless it seems that the majority of patients receiving antipsychotic medication will not exhibit tardive dyskinesia. Once the condition has developed it does not tend to show a steady increase in severity, and may show a fluctuating course with spontaneous remissions (Barnes *et al.* 1983; Casey 1985). Further, tardive dyskinesia is usually mild. Kane *et al.* (1985) found only 8 per cent of their sample of dyskinetic patients had ever warranted a rating of moderately severe.

Disability   Disabilities associated with tardive dyskinesia may be directly related to the movements, such as interference with eating and speech. The condition may also be associated with respiratory irregularity (Yassa and Lal 1986). More generally, patients with this condition may carry an excess burden of mortality and morbidity related to respiratory tract infections and cardiovascular disorders (Youssef and Waddington 1987). However, the oro-facial movements can also represent a social handicap, appearing odd, sometimes grotesque, and a stigma of 'madness' to a casual observer.

Identification of high and low-risk patients

The identification of factors contributing to individual vulnerability to the condition should be of help in its prediction and in the development of strategies to minimize the risk of development. The most important patient variable is undoubtedly age.

Age

Advancing age is the factor most consistently predicting an increased risk of tardive dyskinesia. Examining pooled epidemiological data, Smith and Baldessarini (1980) found a strong linear correlation beween age and both the prevalence and severity of tardive dyskinesia. In most studies, advancing age has been found to be associated with an increased occurrence of tardive dyskinesia, greater severity, and a reduced likelihood of spontaneous remission (Smith and Baldessarini 1980; Barnes *et al.* 1983).

Sex

A relatively consistent finding has been that women show a greater prevalence of severe dyskinesia, although the available evidence suggests that this may only hold true within the geriatric age range.

Brain damage

Early reports on tardive dyskinesia raised the question of an association with brain damage. Several studies have employed a variety of very indirect and tentative measures of brain damage such as evidence of epilepsy or alcoholism or a history of various physical treatments including leucotomy and ECT. Only a few of these studies have revealed an association between brain damage and tardive dyskinesia (see Barnes and Liddle 1985). However, in several surveys of large samples of psychiatric inpatients tardive dyskinesia has been found to be more prevalent in patients with organic mental syndromes such as epilepsy or mental retardation than those with schizophrenia (Hunter *et al*. 1964; Crane and Paulson 1967; Yassa *et al*. 1984).

Further, data from studies employing computerized tomographic brain scans have revealed differences between patients with and without tardive dyskinesia, suggesting more structural damage in those cases with the abnormal movements (Golden *et al*. 1980; Owens *et al*. 1985).

Another relevant measure is the presence of soft neurological signs, such as poor co-ordination. These deficiencies suggest neurological impairment but lack precise cerebral localization. Wegner *et al*. (1985) selected a battery of tests for soft signs that they considered would not be confounded by the presence of dyskinesia. They compared schizophrenic patients with and without tardive dyskinesia and found a positive relationship between the number of soft signs and tardive dyskinesia. Similarly, Youseff and Waddington (1988) reported a strong association between the presence of developmental, or primitive, reflexes and tardive dyskinesia.

Intellectual impairment

In the same study by Wegner and colleagues (1985), schizophrenic patients with dyskinesia performed significantly more poorly on tests of intellectual function, and the balance of evidence from more recent work confirms this association (Barnes and Liddle 1985; Waddington 1987). For example, Struve and Willner (1983) found a positive association between persistent tardive dyskinesia and a reduced ability to think abstractly. The authors interpreted the data from a subgroup of the patients as suggesting that this impairment may have preceded the appearance of the dyskinesia.

Overall, these findings suggest that signs of organic brain damage may be commoner in patients with tardive dyskinesia, and that clinical evidence of brain damage and intellectual or neurological impairment may identify patients with an increased risk of developing tardive dyskinesia with antipsychotic drug treatment.

Drug-induced, early-onset movement disorder

Susceptibility to drug-induced extrapyramidal side-effects as a predictor of tardive dyskinesia was first suggested by Crane (1972). He considered that tardive dyskinesia was more likely to emerge in patients with drug-induced parkinsonism than in patients not developing this early side-effect. A variety of study findings and case-report material suggests that patients presenting with symptoms of parkinsonism, acute akathisia, and acute dystonic reactions are more likely to manifest tardive dyskinesia (Chouinard *et al*. 1979, Barnes and Braude 1984; Kane *et al*. 1985).

Affective illness

Recent work has highlighted affective disturbance as a possible risk factor. Both a positive family history of affective disorder, and the presence of depressive features in schizophrenia, have been mooted as markers of an increased likelihood of developing both parkinsonion side-effects and tardive dyskinesia, but the evidence is tentative at present (see Barnes 1988).

More convincingly, as first noted by Davis and co-workers (1976), a primary diagnosis of affective disorder seems to be associated with a propensity to develop tardive dyskinesia with the administration of anti-psychotic drugs. For example, Mukherjee *et al*. (1986) found persistent tardive dyskinesia in 35 per cent of a sample of bipolar patients who had received maintenance antipsychotic drugs, while no patients without such a drug history had persistent dyskinesia. The preliminary findings of the prospective study by Kane *et al*. (1985) further support psychiatric diagnosis as a relevant variable. A life-table analysis based on the length of drug administration revealed that, after six years of exposure to antipsychotic drugs, the cumulative incidence of tardive dyskinesia in patients with affective or schizo-affective disorder was 26 per cent which was significantly greater than the corresponding figure of 18 per cent for patients with a diagnosis of schizophrenia.

Schizophrenia

The schizophrenic illness itself may also be a risk factor (Barnes and Liddle 1985). Descriptions of abnormal movements similar to tardive dyskinesia appeared in the literature long before the advent of antipsychotic drugs, suggesting that schizophrenia is associated with a tendency to develop abnormal involuntary movements. Owens (1985) has propounded the notion that antipsychotic drugs act to promote or exacerbate a tendency to develop spontaneous movement disorders inherent in at least some forms of the illness. He has suggested that tardive dyskinesia is part of the type II syndrome of schizophrenia.

If this were true, one would expect tardive dyskinesia to be more common in patients with negative features of schizophrenia. Reviews of studies that have examined this issue seem to support this view (Barnes and Liddle 1985; Waddington *et al*. 1987). A consistent finding has been a significant association between tardive dyskinesia and blunting of affect. Tardive dyskinesia has also been found to be variously associated with other negative schizophrenic features including social and emotional withdrawal, psycho-motor retardation, self-neglect, and muteness. In our own study (Barnes *et al*.

1989) examination of the association between age and tardive dyskinesia in patients with and without negative symptoms suggested that patients with such symptoms were more susceptible to the early development of tardive dyskinesia.

However, these findings within groups of schizophrenic patients should be seen in context. Commenting on the issue of psychiatric diagnosis as a risk factor, Casey and Keepers (1988) conclude that schizophrenic patients may be less vulnerable to tardive dyskinesia when compared with either patients who have other psychiatric disorders or individuals without psychiatric illness exposed to antipsychotics.

Summary

These patient variables, identified with varying degrees of certainty as risk factors for tardive dyskinesia, may allow for some limited assessment of the risk of tardive dyskinesia in individuals starting antipsychotic drug treatment. For example, when weighing up the potential risks and benefits of such treatment for patients who are elderly, brain-damaged, or suffering from affective disorder, the increased likelihood of tardive dyskinesia in these patients should be taken into account.

Prophylactic antipsychotic treatment regimes

The majority of controlled investigations have failed to discover a relationship between increased risk of tardive dyskinesia and the total amount of antipsychotic drug received, the current dosage or the duration of drug therapy. However, Baldessarini *et al.* (1988) have speculated that this absence of convincing epidemiological evidence for a dose–risk relationship is due to the excessive doses commonly administered, which overwhelm and mask any such association. They concluded from their review of the relevant literature that doses of drug above 600 mg chlorpromazine equivalents per day are not only more likely to be associated with side-effects but also, perhaps, with a reduction in clinical benefit. Strategies aimed at reducing the amount of drug administered long term include so-called 'drug holidays', low-dose treatment, and brief intermittent or 'targetted' therapy.

Drug holidays

Drug holidays for patients on long-term antipsychotic treatment were first advocated as a beneficial strategy by Ayd (1966). He suggested that intermittent therapy might reduce the risk of developing tardive dyskinesia. The American College of Neuropsychopharmacology—Food and Drug Administration Task Force (1973) supported this notion. However, the findings of clinical studies that have investigated this treatment variable have suggested that rather than reduce the likelihood of tardive dyskinesia, interruptions of drug therapy may increase the risk of persistent dyskinesia as well as the risk of relapse (Goldman and Luchins 1984).

Low-dose studies

Information on the dose–response relationship for tardive dyskinesia is scant. Patients vulnerable to the development of the condition, perhaps on the basis of some of the risk factors already described, may develop dyskinesia on relatively modest doses of antipsychotic medication while those without such a susceptibility may be at low risk of dyskinesia even when receiving doses above the normal range. However, data available from controlled studies suggest that low-dose antipsychotic treatment may be associated with a reduced incidence of early signs of tardive dyskinesia (Kane *et al.* 1983) as well as improved social functioning and sense of well-being. Unfortunately, the characteristics of schizophrenic patients who are likely to respond to low-dose drug therapy are not known. The principal risk is an increased tendency to relapse, but this may not be apparent for several years if the reduction in dose is modest (Johnson *et al.* 1987).

Brief intermittent treatment

Similar considerations apply to what has been called 'brief intermittent' or 'targeted' treatment, pioneered by Herz and Melville (1980). Maintenance medication is discontinued but there is prompt, though time-limited, reinstitution of antipsychotic drug treatment when early signs of relapse appear. Patients and their relatives learn to recognize these early or 'prodromal' signs. Close clinical surveillance of patients is required. In the short term, there are claims for fewer signs of tardive dyskinesia with this treatment strategy (Herz 1988; Jolley *et al.* 1989) but the results from follow-up over several years are required to evaluate the viability and clinical advantages and disadvantages of this treatment approach.

Anticholinergic agents

The administration of anticholinergic agents will invariably exacerbate existing tardive dyskinesia, or it may provoke the appearance of the condition. Withdrawal of these drugs is also likely to improve tardive dyskinesia. However, the popular view that the chronic prescription of anticholinergics increases the risk of tardive dyskinesia is not confirmed by the available evidence. Klawans *et al.* (1980) reviewed the literature and stated that an increased prevalence of tardive dyskinesia in patients receiving chronic anticholinergic drugs had not been established. Similarly, Gardos and Cole (1983) concluded that anticholinergics could not be considered a risk factor for tardive dyskinesia.

Identification of antipsychotic drugs with a low liability for tardive dyskinesia

There is no strong evidence that any antipsychotic drug has a lower liability to cause tardive dyskinesia. Retrospective studies are confounded by factors such as polypharmacy, uncertainties about compliance with medication, and the administration of other drugs. The long-term comparative drug studies that would be necessary, involving patients starting and remaining on a single antipsychotic drug over several years, have not been undertaken and are probably impractical. The theoretical basis on which to predict the relative risk of tardive dyskinesia with a particular drug or class of drugs is uncertain, but components of the pharmacological activity and certain clinical effects have been considered as possibly being of relevance.

Principally on the basis of rodent studies, several pharmacological properties have been considered relevant to the ability of a particular antipsychotic drug to induce tardive dyskinesia. These include the relative affinities for D-1 and D-2 dopamine receptors, the ability to induce D-2 dopamine receptor supersensitivity, and a selective action on brain dopaminergic systems (Jenner *et al.* 1985; Coward *et al.* 1988).

Clinical effects that might provide some indication of the relative risk for tardive dyskinesia with a drug could include the relative severity of dyskinesia provoked or exacerbated by withdrawal of the drug, and the incidence figures for patients maintained on the drug alone. The ability to suppress tardive dyskinesia is common to virtually all antipsychotic drugs and is therefore unlikely to be of predictive value, although the relative rapidity with which the dyskinesia reappeared might be of interest.

Sulpiride

The strongest claims for a reduced liability for tardive dyskinesia have been made for sulpiride and clozapine, and for both drugs various theories have been put forward to explain this potential advantage. Sulpiride is a substituted benzamide, and a selective antagonist of D-2 dopamine receptors, and possibly a subgroup of these, with no significant effects on other types of receptor so far tested. Striatal dopaminergic supersensitivity is commonly held to be the pathophysiological basis of tardive dyskinesia. In animal studies, chronic administration of haloperidol causes a dose-dependent increase in receptor supersensitivity in the striatal dopamine system while similar treatment with sulpiride fails to induce supersensitivity (Rupniak *et al.* 1984). Uncontrolled worldwide observations reveal only a relatively small number of patients receiving sulpiride who have developed tardive dyskinesia; however, claims for a lower risk of the condition remain to be substantiated.

Clozapine

Three long-term retrospective studies of patients treated with clozapine revealed no patients receiving clozapine alone who had developed tardive dyskinesia. From his review of the available evidence, Casey (1988) also concluded that thus far the drug has a very low incidence of parkinsonism, and there are virtually no reports of akathisia or dystonia. Marder and Van Putten (1988) concur that there are no convincing reports of tardive dyskinesia associated with clozapine, but point out that this potential advantage of clozapine remains tentative in the absence of adequate incidence data in patients receiving this drug. Also, the drug is associated with a relatively high rate of agranulocytosis. A recent study of clozapine found a cumulative incidence of 2 per cent for this toxic effect after a year of treatment (Kane *et al.* 1988). Assiduous haematological monitoring is therefore necessary.

Explanations put forward for the apparent lack of tardive dyskinesia with clozapine refer to pharmacological differences from the typical or standard antipsychotics. The drugs causes a preferential blockage of D-1 dopamine receptors rather than D-2 receptors, and leads to an increase rather than decrease in GABA turnover. Also, like sulpiride, animal work suggests that clozapine does not induce dopamine receptor supersensitivity (Coward *et al.* 1988).

Whether the relative selectivity of drugs for D-1 and D-2 receptors is relevant to their ability to induce tardive dyskinesia remains unclear (Kistrup and Gerlach 1987). For example, from their own rodent studies, Jenner *et al.* (1985) considered that the low incidence of tardive dyskinesia claimed for both clozapine and sulpiride was related to the absence of change in D-2 function with chronic administration, coupled with an ability to enhance D-1 function. However, this conclusion seems at odds with a report by Ellison *et al.* (1988) that oral movements in rodents associated with chronic administration of antipsychotic drugs can be more clearly induced by a D-1 agonist than a D-2 agonist. The necessary extrapolation of such findings from animals to humans increases the difficulty of interpreting the clinical relevance of the findings from these and other similar studies in this area.

Summary

The apparently lower risk of tardive dyskinesia with clozapine and sulpiride raises the hope that effective antipsychotic drugs can be developed in the future which do not have the capacity to induce this condition. However, as with clozapine, this advantage may be offset by the risk of other serious side-effects. Similarly, low-dose drug treatment offers a simple clinical strategy with claims for a reduced incidence of tardive dyskinesia, but the risk of relapse is increased. At present, the information available on which to judge the risk–benefit ratio of this or any other drug regime in an individual case is inadequate in respect of therapeutic efficacy on the one hand and side-effects such as tardive dyskinesia on the other.

Drug treatment

This chapter has focused on possible preventive measures rather than the treatment of tardive dyskinesia once it has appeared. Tetrabenazine is perhaps the drug with the best evidence for suppression of dyskinesia (*Drug and Therapeutics Bulletin* 1986), but the list of putative remedies continues to grow, with recent claims for calcium channel blockers, such as verapamil and diltiazem (Barrow and Childs 1986; Ross *et al.* 1987) and vitamin E (Lohr *et al.* 1987), among others. There is a respectable amount of evidence for GABA agonists such as sodium valproate and progabide as antidyskinetic drugs with a beneficial effect on tardive dyskinesia. Tamminga and Thaker (1989) suggest a possible prophylactic action for these agents but this intriguing speculation remains to be tested.

Clinical guidelines

1. The core feature of tardive dyskinesia is oro-facial dyskinesia. Choreiform trunk and limb movements may constitute a separate subsyndrome, and an attempt should be made to distinguish this from tardive dystonia, tardive akathisia, and schizophrenic movement disorder such as stereotypies and posturing. This distinction is often extremely difficult, but it is pertinent to management and the assessment of response to treatment.

2. Consideration of the known risk factors for tardive dyskinesia (increasing age, female sex, brain damage, intellectual impairment, early-onset drug-induced movement disorders, affective disorder, and negative symptoms of schizophrenia) may allow identification of at least some patients with an increased likelihood of developing the condition with antipsychotic drug treatment.

3. Withdrawal of anticholinergic agents will usually improve tardive dyskinesia but in each case this benefit must be balanced against the re-emergence of parkinsonism.

4. There is no convincing evidence for a lower liability for tardive dyskinesia with any particular antipsychotic drug or class of drug, although such claims have been made for sulpiride and clozapine.

5. The degree to which low-dose and intermittent targeted drug treatment may reduce the incidence of tardive dyskinesia remains unclear and must be weighed against an increased risk of relapse with these drug strategies.

6. Recommendations for the use of antipsychotic drugs may be summarized as, first, to prescribe drugs only when clearly indicated, and continue administration only where there is evidence of benefit. The risk–benefit ratio for continued drug therapy may warrant particular consideration in patients with affective disorder. Second, antipsychotic drugs should be used judiciously, with maintenance medication at the minimum effective dose. Third, interruptions of drug treatment should be avoided.

References

American College of Neuropsychopharmacology—Food and Drug Administration Task Force (1973). Neurological syndromes associated with antipsychotic drug use. *New England Journal of Medicine,* **289,** 20–3.

Ayd, F. J. (1966). Drug holidays: intermittent pharmacotherapy for psychiatric patients. *International Drug Therapy News Letter,* **8,** 1–3.

Baldessarini, R. J., Cohen, B. M., and Teicher, M. H. (1988). Significance of neuroleptic dose and plasma level in the pharmacological treatment of psychosis. *Archives of General Psychiatry,* **45,** 79–91.

Barnes, T. R. E. (1988). Tardive dyskinesia: risk factors, pathophysiology and treatment. In *Recent advances in clinical psychiatry, volume 6* (ed. K. Granville-Grossman), pp. 185–207. Churchill Livingstone, London.

Barnes, T. R. E. and Braude, W. M. (1984). Persistent akathisia associated with early tardive dyskinesia. *Postgraduate Medical Journal,* **60,** 51–3.

Barnes, T. R. E. and Liddle, P. F. (1985). Tardive dyskinesia: implications for schizophrenia? In *Schizophrenia: new pharmacological and clinical development* (ed. A. A. Schiff, Sir Martin Roth, and H. L. Freeman), pp. 81–7. Royal Society of Medicine Services, London.

Barnes, T. R. E., Kidger, T., and Gore, S. M. (1983). Tardive dyskinesia: a 3-year follow-up study. *Psychological Medicine,* **13,** 71–81.

Barnes, T. R. E., Liddle, P. F., Curson, D. A., and Patel, M. (1989). Negative symptoms, tardive dyskinesia and depression in chronic schizophrenia. *British Journal of Psychiatry,* Supplement **7,** 99–103.

Barrow, N. and Childs, A. (1986). An anti-tardive dyskinesia effect of verapamil. *American Journal of Psychiatry,* **143,** 1485.

Blowers, A. J., Borison, R. L., Blowers, C. M., and Bicknell, D. J. (1982). Abnormal involuntary movements in the elderly. *British Journal of Psychiatry,* **139,** 363–4.

Casey, D. E. (1985). Tardive dyskinesia: reversible and irreversible. In *Dyskinesia—research and treatment* (ed. D. E. Casey, T. N. Chase, A. V. Christensen, and Gerlach, J.), pp. 88–97. Springer-Verlag, Berlin.

Casey, D. E. (1988). Neuroleptic-induced EPS and TD. Paper presented at Clozapine (Leponex/Clozaril) Scientific Update meeting, 31 October–1 November 1988, Montreux, Switzerland.

Casey, D. E. and Keepers, G. A. (1988). Neuroleptic side effects: acute extrapyramidal syndromes and tardive dyskinesia. In *Psychopharmacology: current trends* (ed. D. E. Casey and A. V. Christensen), pp. 74–93. Springer-Verlag, Berlin.

Chouinard, G., Annable, L., Ross-Chouinard, A., and Nestoros, J. N. (1979). Factors related to tardive dyskinesia. *American Journal of Psychiatry*, **136**, 79–83.

Coward, D. M., Imperato, A., Urwyler, S., and White, T. G. (1988). Biochemical and behavioural properties of clozapine. Paper presented at Clozapine (Leponex/Clozaril) Scientific Update meeting, 31 October–1 November 1988, Montreux, Switzerland.

Crane, G. E. (1972). Pseudoparkinsonism and tardive dyskinesia. *Archives of Neurology*, **27**, 426–30.

Crane, G. E. and Paulson, G. (1967). Involuntary movements in a sample of chronic mental patients and their relation to the treatment with neuroleptics. *International Journal of Neurology*, **3**, 286–91.

Davis, K., Berger, P., and Hollister, L. (1976). Tardive dyskinesia and depressive illness. *Psychopharmacology Communications*, **2**, 125.

Drug and Therapeutics Bulletin (1986). The management of tardive dyskinesia. *Drug and Therapeutics Bulletin*, **24**, 27–8.

Ellison, G., Johansson, P., Levin, E., See, R., and Gunne, L. (1988). Chronic neuroleptics alter the effects of the D1 agonist SK&F 38393 and the D2 agonist LY 171555 on oral movements in rats. *Psychopharmacology*, **96**, 253–7.

Gardos, G. and Cole, J. O. (1983). Tardive dyskinesia and anticholinergic drugs. *American Journal of Psychiatry*, **140**, 200–2.

Golden, C. J., Moses, J. A., Zelazowski, R., Graber, B., Zatz, L. M., Horvath, T. B., and Berger, P. A. (1980). Cerebral ventricular size and neuropsychological impairment in chronic schizophrenics. *Archives of General Psychiatry*, **37**, 619–23.

Goldman, M. B. and Luchins, D. J. (1984). Intermittent neuroleptic therapy and tardive dyskinesia: a literature review. *Hospital and Community Psychiatry*, **35**, 1215–19.

Herz, M. I. (1988). Intermittent versus maintenance medication. Paper presented at the Fourth Bi-Ennial Winter Workshop on Schizophrenia. 24–29 January 1988, Badgastein, Austria.

Herz, M. I. and Melville, C. (1980). Relapse in schizophrenia. *American Journal of Psychiatry*, **137**, 801–5.

Hunter, R., Earl, C. J., and Thornicroft, S. (1964). An apparently irreversible syndrome of abnormal movements following phenothiazine medication. *Proceedings of the Royal Society of Medicine*, **57**, 758.

Jenner, P., Rupniak, N. M. J., and Marsden, C. D. (1985). Differential alteration of striatal D-1 and D-2 receptors induced by the long-term administration of haloperidol, sulpiride or clozapine to rats. In *Dyskinesia—research and treatment* (ed. D. E. Casey, T. N. Chase, A. V. Christensen, and Gerlach, J.), pp. 174–81. Springer-Verlag, Berlin.

Johnson, D. A. W., Ludlow, J. M., Street, K., and Taylor, R. D. W. (1987). Double-blind comparison of half-dose and standard-dose flupenthixol decanoate in the maintenance treatment of stabilised outpatients with schizophrenia. *British Journal of Psychiatry*, **151**, 634–8.

Jolley, A. G., Hirsch, S. R., McRink, A., and Manchanda, R. (1989). Trial of brief intermittent neuroleptic prophylaxis for selected schizophrenic outpatients: clinical outcome at one year follow-up. *British Medical Journal*, **298**, 985–90.

Kane, J. M. and Smith, J. M. (1982). Tardive dyskinesia: prevalence and risk factors, 1959–1979. *Archives of General Psychiatry*, **39**, 473–81.

Kane, J. M., Weinhold, P., Kinon, B., Wegner, J., and Leader, M. (1982). Prevalence of abnormal involuntary movements ('spontaneous dyskinesia') in the normal elderly. *Psychopharmacology*, **77**, 105–8.

Kane, J. M., Rifkin, A., Woerner, M., Reardow, G., Sarantakos, S., Schiebel, D., and Ramos-Lorenzi, J. (1983). Low-dose neuroleptic treatment of outpatient schizophrenics. *Archives of General Psychiatry*, **40**, 893–6.

Kane, J. M., Woerner, M., and Leiberman, J. (1985). Tardive dyskinesia: prevalence, incidence and risk factors. In *Dyskinesia—research and treatment* (ed. D. E. Casey, T. N. Chase, A. V. Christensen, and Gerlach, J.), pp. 72–8. Springer-Verlag, Berlin.

Kane, J. M., Honigfeld, G., Singer, J., Meltzer, H., and the Clozaril Collaborative Study Group (1988). Clozapine for the treatment-resistant schizophrenic. *Archives of General Psychiatry*, **45**, 789–96.

Kistrup, K. and Gerlach, J. (1987). Selective D1 and D2 receptor manipulation in cebus monkeys: relevance for dystonia and dyskinesia in humans. *Pharmacology and Toxicology*, **61**, 157–61.

Klawans, H. L., Goetz, C. G., and Perlik, S. (1980). Tardive dyskinesia, review and update. *American Journal of Psychiatry*, **137**, 900–8.

Lohr, J. B., Cadet, J. L., Lohr, M. A., Jeste, D. V., and Wyatt, R. J. (1987). Alpha-tocopherol in tardive dyskinesia. *Lancet*, **1**, 913–14.

Marder, S. R. and Van Putten, T. (1988). Who should receive clozapine? *Archives of General Psychiatry*, **45**, 865–7.

Mukherjee, S., Rosen, A. M., Caracci, G., and Shukla, S. (1986). Persistent tardive dyskinesia in bipolar patients. *Archives of General Psychiatry*, **43**, 342–6.

Owens, D. G. C. (1985). Involuntary disorders of movement in chronic schizophrenia—the role of the illness and its treatment. In *Dyskinesia—research and treatment* (ed. D. E. Casey, T. N. Chase, A. V. Christensen, and Gerlach, J.), pp. 79–87. Springer-Verlag, Berlin.

Owens, D. G. C., Johnstone, E. C., Crow, T. J., Frith, C. D., Jagoe, J. R., and Kreel, L. (1985). Lateral ventricular size in schizophrenia: relationship to the disease process and its clinical manifestations. *Psychological Medicine*, **15**, 27–41.

Ross, J. L., Mackenzie, T. B., Hanson, D. R., and Charles, C. R. (1987). Diltiazem for tardive dyskinesia. *Lancet*, **1**, 268.

Rupniak, N. M. J., Mann, S., Hall, M. D., Fleminger, S., Kilpatrick, G., Jenner, P., and Marsden, C. D. (1984). Differential effects of continuous administrations for 1 year of haloperidol or sulpiride on striatal dopamine function in the rat. *Psychopharmacology*, **84**, 503–11.

Schonecker, M. (1957). Beitrag zu der Mitteilung von Kulenkampff und Tarnow: Ein eigentumliches Syndrom im oralen Bereich bei Megaphenapplikation. *Nervenarzt*, **28**, 35.

Smith, J. M. and Baldessarini, R. J. (1980). Changes in prevalence, severity and recovery in tardive dyskinesia with age. *Archives of General Psychiatry*, **37**, 1368–73.

Struve, F. A. and Willner, A. E. (1983). Cognitive dysfunction and tardive dyskinesia. *British Journal of Psychiatry*, **143**, 597–600.

Tamminga, C. A. and Thaker, G. K. (1989). Tardive dyskinesia. *Current Opinion in Psychiatry*, **2**, 2–16.

Toenniessen, L. M., Casey, D. E., and McFarland, B. H. (1985). Tardive dyskinesia in the aged. *Archives of General Psychiatry*, **42**, 278–284.

Waddington, J. L. (1987). Tardive dyskinesia in schizophrenia and other disorders: association with ageing, cognitive dysfunction and structural brain pathology in relation to neuroleptic exposure. *Human Psychopharmacology*, **2**, 11–22.

Waddington, J. L., Youssef, H. A., Dolphin, C., and Kinsella, A. (1987). Cognitive dysfunction, negative symptoms and tardive dyskinesia in shizophrenia. Their association in relation to topography of involuntary movements and criteria of their abnormality. *Archives of General Psychiatry*, **44**, 901–12.

Wegner, J. T., Catalano, F., Gibralter, J., and Kane, J. M. (1985). Schizophrenics with tardive dyskinesia, neuropsychological deficit and family psychopathology. *Archives of General Psychiatry*, **42**, 860–5.

Yassa, R. and Lal, S. (1986). Respiratory irregularity and tardive dyskinesia. *Acta Psychiatrica Scandinavica*, **73**, 506–10.

Yassa, R., Nair, B., and Schwartz, G. (1984). Tardive dyskinesia and the primary psychiatric diagnosis. *Psychosomatics*, **25**, 135–8.

Youssef, H. A. and Waddington, J. L. (1987). Morbidity and mortality in tardive dyskinesia: associations in chronic schizophrenia. *Acta Psychiatrica Scandinavica*, **75**, 74–7.

Youssef, H. A. and Waddington, J. L. (1988). Primitive (developmental) reflexes and diffuse cerebral dysfunction in schizophrenia and bipolar affective disorder: overrepresentation in patients with tardive dyskinesia. *Biological Psychiatry*, **23**, 791–6.

14

Alcohol dependence: what treatment works?

JONATHAN CHICK

As with other conditions which present to doctors and which have neither unitary causation nor an established cure, treatments for alcohol dependence abound and controversy is rife. An International Congress in 1988 contained a paper from the Soviet Union comparing 19 forms of treatment including camphoric-turpentine baths versus intranasal lithium electrophoresis. If our discussion is restricted to the English-speaking world, the field of potential controversy becomes substantially simplified. There is a fair consensus about what constitutes the core of treatment: encouraging the person to take responsibility for his or her drinking, to find other ways of 'coping with his problems', and to 'learn to live without alcohol' (or with very little); in short, a psychological approach.

Apart from vitamins and tranquillizers at the point of detoxification, there has been little attention given to physical approaches to helping people stay well. Some important recent studies have hinted at benefit in the first few months of treatment of non-addictive medication such as lithium (Fawcett et al. 1987), the dopamine antagonist tiapride (Shaw et al. 1987), or 5-HT uptake inhibitors (Naranjo et al. 1986). There is an issue here about whether there is a 'sub-acute withdrawal syndrome', that is, some chemical lack, which goes beyond the recognized 7–14 day duration of detoxification treatment and which makes it particularly difficult for patients to control their drinking in the first six months or so. Or it may be that lithium and 5-HT uptake inhibitors reduce some of the positive reinforcing properties of alcohol (without actually providing an aversive reaction).

There is a feeling among some recovered patients that alcoholics should not take drugs for their condition, but this has not led to much argument in the literature. No, the controversy that has generated most heat has not been physical versus psychological treatment, or psychodynamic versus behavioural psychotherapy. It has not even been the grand theoretical issue of the relative causal contributions of the individual biochemistry of alcohol-metabolism versus learning and social forces (though perhaps this has hovered in the background). It has been what, to an outsider, may seem curiously particulate. It is whether therapists should ever countenance any attempt by the patient to continue some drinking or resume limited drinking of alcohol, or whether they should insist that total abstinence is the only route to well-being.

Abstinence or controlled drinking?

To many it seems self-evident that the only route to health and sanity for a person whose life has been severely affected by alcohol is through total abstinence. The challenge to such absolutism (e.g., Davies (1962) and the Sobells' comparison of controlled drinking versus abstinence-oriented therapy (1976)) raised a storm which even went far beyond the alcoholism treatment world, with newspaper editorials, Committees of Inquiry, and allegations in the prestigious journal *Science* which implied that some researchers who had claimed that controlled drinking was possible had faked results.

To an extent, the earlier opponents of abstinence absolutism set up a straw man by implying that the abstinence theory stated that all people dependent on alcohol experience irresistible craving as soon as they take one drink, whatever the circumstances. Clinical experience as well as research has not surprisingly refuted such an extreme position (Chick 1986; Nordstrom and Berglund 1987). Powerful therapies often necessarily simplify matters to provide easily grasped concepts for new adherents, but Alcoholics Anonymous' teaching, as I understand it, is not that 'one drink, one drunk' means an inevitable physiological tripwire. Rather, the point is that the severely addicted drinker can no longer predict with certainty that if he resumes drinking it will not sooner or later escalate.

The research tends to show that the problem drinkers in the community who are most likely to achieve controlled drinking are young, have had fewer problems, a short heavy-drinking history and are not severely dependent. It is as if, provided the drinker pulls himself up in time, certain malignant physiological or learnt processes do not develop, and control of drinking is more easily regained.

Whether controlled drinking can be taught to clinic attenders is a separate question. Some will intermittently succeed in sticking to a limit and will reduce over the years the total harm caused by their drinking. But a recent, well-designed study in severely dependent patients found evidence that such patients do better at least in the first six months with abstinence-oriented treatment than with training aimed at controlled drinking (though the advantage to the abstinence-oriented therapy was no longer visible at the 5–6 year follow-up) (Rychtarik *et al.* 1987).

Nevertheless, there are still alcohol-dependent drinkers, who do not have serious physical damage from alcohol, and who want to resume drinking and believe they can succeed. We should stick with such a patient and try to help him or her find ways of resuming drinking with minimum harm. This should be done in the spirit of an experiment whose results are to be objectively validated, for example, with reports from third parties such as family or employer, and blood tests (serum gamma glutamyl transferase and mean red-cell volume). If the experiment fails, the patient is still in touch with the therapist so that the strategy can be modified towards complete abstinence without giving up the venture altogether. Success in controlled drinking therapy is greater if commenced after at least a few weeks of abstinence.

Controlled drinking therapy is most likely to succeed in early-stage cases (Sanchez-Craig *et al.* 1984). Early-stage cases are those who have not regularly had morning tremor, do not report regularly having days or being in situations when they fell restless without a drink, or are preoccupied by the thought of getting it, and their drinking has not repeatedly escalated after attempts to cut down. They are under 40, or else have been having problems from their drinking for, say, less than three years. I suggest also that problem-free drinking is most likely to be achieved by patients whose personality profile suggests internal controls and lack of impulsivity; who are socially stable; whose spouse agrees with the plan; (and, perhaps, who agree to remain in occasional contact with the therapist).

There may be few such cases attending psychiatric clinics. But there will certainly be some for whom cutting down is all that is needed among the early problem drinkers identified in general practice and general hospitals, during employee health checks, or picked up via health-screening projects or newspaper advertisements. And it is in these settings that controlled studies to date show consistently that an intervention motivating the patient and offering advice can be effective (see below).

However, severely alcohol-dependent patients should be advised to abstain and only if this goal is rejected by the patient or if it repeatedly fails should controlled drinking be worked at for an experimental period. This was also the majority view in a recently published set of responses to this question solicited by the *British Journal of Addiction* (1987).

Use of deterrent medication

For those drinkers who state a wish for abstinence, the regular use of disulfiram (Antabuse) (or calcium carbimide, Abstem) may assist them to achieve that aim. The doctor explains to patients opting for this aid that an unpleasant or even dangerous reaction will occur if the patient drinks alcohol. This deters many, though not all, patients from drinking for as long as they take the tablets.

The most recent and thorough study of the use of disulfiram was a multi-centre one-year follow-up of 605 men randomly assigned to either 250 mg of disulfiram (a usual dosage), 1 mg disulfiram (insufficient to cause an alcohol reaction), or simply riboflavin (Fuller *et al.* 1986). Riboflavin was used as a marker in the disulfiram tablets so that an estimate of compliance could be obtained. About a fifth of patients complied adequately, and irrespective of the medication they had been allocated to, nearly half of these achieved a year's abstinence. Of non-compliers, only 8 per cent were abstinent for a year. In 47 per cent of the 438 patients who reported some drinking, sufficient data were available on patterns of consumption to establish that, in this group at least, allocation to 250 mg disulfiram was associated with significantly fewer drinking days.

Are there risks of toxicity?

Considering that alcoholism treatment results obtained by psychiatrists are far from encouraging, it surprises me that deterrent drugs are not more often offered to patients. One reason is apprehension about toxic effects. In the 40 years that disulfiram has been in use, there have been a small number of reports of fatal disulfiram-ethanol reactions and of serious liver toxicity. But these cases are extremely rare and today the drug is not prescribed to patients with liver or heart disease. Minor side-effects are also rare, but occasionally patients develop tiredness and headache which usually responds to a reduction of the dose. Bad breath was the only complaint that was more common in the disulfiram group in a double-blind, placebo-controlled study of side-effects in 158 Danish patients (Christensen *et al.* 1984). However, peripheral neuropathy has been reported with disulfiram. It disappears once the drug is withdrawn. Calcium carbimide seems to have fewer side-effects, but like disulfiram, skin allergies are sometimes seen.

Calcium carbimide only offers about 24 hours' risk of an ethanol-reaction, whereas disulfiram, if taken regularly, leads to a reaction up to 5–7 days after the last dose, and is thus the preferred deterrent drug. However, change to calcium carbimide is indicated if allergy to disulfiram develops. Disulfiram slows the metabolism of some drugs which may be prescribed in psychiatry, such as phenytoin and tricyclics, but this simply means lowering the dose appropriately. Disulfiram should not be prescribed to schizo-phrenics, in whom it may lead to psychotic symptoms.

In my view, the morbidity and mortality of severely alcohol-dependent patients is sufficiently high to justify offering deterrent treatment, despite the small risk of side-effects, to many patients. All patients should, however, be fully informed about the drug and the risks, as well as the alcohol reaction, preferably in a written information sheet in addition to their discussion with the doctor.

Supervised disulfiram

Another reason why deterrent drugs fell out of use was that doctors found that patients stopped taking the tablets, or took them sporadically so that no reaction resulted from drinking (disulfiram only causes an alcohol-aversive reaction if it has been taken for some 5–7 days). When attempts are made to ensure that disulfiram is taken regularly, for example, by an agreement with the spouse, its advantages are clearer, even striking, as in Azrin's studies (e.g., Azrin 1976).

In several studies, the taking of disulfiram under supervision as part of a court requirement has been shown to improve outcome and prevent re-cidivism in otherwise often recalcitrant alcohol abusers such as drink-drive offenders and drunken offenders (Chick 1990). Robichaud *et al.* (1979) specifically studied a group who had failed on many occasions to respond to treatment and who had had their final warning at their workplace. These employees were offered a final chance to keep their jobs provided they took disulfiram supervised by the clinic. Most men accepted this, and their annual absenteeism rate fell from a mean of 9.8 per cent in the year prior to the treatment to 1.7 per cent during the disulfiram year. Many doctors who run alcoholism clinics weary of that quota of regularly readmitted relap-sers. Sereny *et al.* (1986) told all such patients during a given period that they would have to accept clinic-supervised disulfiram if they wished to continue a relationship with the clinic. Apparently 68 of 73 did so. There was no control group in this study but a remarkable proportion of these relapsers remained sober for such long periods while on supervised dis-ulfiram that many were able to be discharged from the clinic after 12 months' sobriety.

In my view, British alcoholism treatment has been excessively cautious in offering supervised disulfiram to patients who would otherwise fail to respond to treatment and perhaps be rejected by busy clinics, their failure to stop drinking eventually being tautologously ascribed to 'poor motiva-tion'. There is a misplaced fear that the patient is being wrongly coerced. This reflects a general reluctance to accept coerced patients into treatment (ignoring that fact that most alcoholics come into treatment under some sort of external pressure such as pressure from the employer or spouse). This reluctance is not justified by the published literature which shows that in many types of mandatory treatment the attendance at treatment and the outcome is often equal to that of voluntary patients (Chick 1990).

The controversy over intensive versus minimal treatment

When recovered problem drinkers identified in community surveys are asked about the cause of their remission, they by no means always mention treatment. In the 1940s, 456 males from disadvantaged Boston neighbour-hoods were studied in depth and followed at intervals till the age of 47, with contact with relatives. Criteria for alcohol abuse at some time in their lives had been fulfilled in 120 men. By age 47, 10 per cent of the alcohol abusers had died. Of the survivors, 21 were securely abstinent at age 47, but of these only four had had treatment in a clinic or hospital, though seven were attenders of Alcoholics Anonymous (AA) (Vaillant *et al.* 1983). In another follow-up study, the Social Research Group in California were able to comment on the recovery pattern in 59 men who resembled clinic attenders in the numbers of alcohol-related problems each admitted to, but who had never in their lives had treatment. Four years later, 14 per cent were free of problems and a further 18 per cent had only minimal problems, but no treatment had been received. The rate of good outcome, i.e., about one-third, is comparable with four-year follow-up studies of clinic attenders, again indicating that recovery without treatment is commonplace (for details see Chick 1982).

Studies of minimal treatment

The controlled study, where attenders at clinics are randomly allocated to no or minimal treatment versus 'optimal' treatment, is the correct approach to learning what difference treatment makes. Studies tend to show that minimal treatments are associated with as good results as intensive treatments. This was the result when out-patient treatment was compared to in-patient treatment (Edwards and Guthrie 1967); and when day care was compared to in-patient care (McLachlan and Stein 1982; Potamianos *et al.* 1986; McCrady *et al.* 1986). The same result was obtained by Edwards *et al.* (1977) even when just one session of advice was compared to a package of comprehensive treatment including group and individual therapy and often admission. They studied only married male patients. Subjects randomly allocated to the advice group only spent one morning at the out-patient clinic. After the research assessment, they had about an hour with an empathic and authoritative therapist who advised them to abstain totally from alcohol, and gave some marital counselling. Thereafter the only contact was between the research team and the patient's wife (which perhaps in retrospect was not an insignificant type of therapy). Some of the men in the advice-only group did obtain treatment at other centres in the following year. Although there was 'a hint that a small group of specially keen AA attenders may have attained a better outcome', the authors concluded that the improvement, which occurred in the 'minimal' group with the same frequency as in the 'optimally treated' group, could not be ascribed in general to the help sought elsewhere.

In a replication and extension of that study, Chick *et al.* (1987) also compared very minimal treatment with a broad package. The 'intensive package' included some in-patient and group therapy and systematic follow-up for those patients who accepted it, but did not include a vigorous Alcoholics Anonymous education nor much emphasis on deterrent drugs. The sample included women and single people, rather than just married men. The research showed that although the stable abstinence rate was no higher after two years in the more intensively treated group, the 'intensive' group had sustained less in terms of total problems related to their drinking, particularly in the family realm, than the 'advice-only' group.

The Minnesota method

In the Western world, the intensive four-plus weeks in-patient treatment based on group therapy and the Alcoholics Anonymous (AA) approach probably consumes by far the largest proportion of expenditure on alcoholism treatment. The model is sometimes known as the Hazelden or Minnesota model (Cook 1988*a*). Recovered alcoholics (AA members) assist in the treatment, which hinges around group confrontation aimed at helping patients accept the reality of the harm their drinking has caused, and the importance of accepting that they have an illness, i.e., alcoholism. That so much should be spent on this form of treatment has been cogently criticized by Miller (e.g., 1987), chiefly on the grounds that no controlled outcome study has demonstrated the model's efficacy (e.g., Cook 1988*b*). Since Cook's paper, a single relevant study has been published from Finland (Keso 1988). In terms of self-reported abstinence rates, Keso found a Hazelden-type programme to be superior to a rather dilute and protracted psychiatric programme. There was no superiority in terms of objective markers of consumption (blood tests) and the study's value is diminished because data from relatives or other third parties was not obtained. Further studies, with randomization to a control group and thorough independent follow-up are required if this type of treatment is to survive the growing worries of governments and insurance companies who face the expense. The value and importance of Alcoholics Anonymous as a support group and route to sobriety for some patients is a separate question and, in my opinion, not controversial. All alcoholism therapists know patients who have benefited from AA; and AA asks for no public funds.

Behaviour therapy approaches

Particularly in early milder cases, therapies based on behavioural and cognitive theory such as training in 'problem-solving skills', training in handling precipitants to relapse, anxiety management, and social skills education, have a more hopeful research literature than some other types of 'intensive' treatment (see Miller 1987 for references up to 1986, and Eriksen *et al.* 1986 for a subsequent and nicely designed, if small, study).

There are few other hints of what aspects of the intensive programme may be important. Costello's meta-analysis (1980) showed that programmes which involve other members of the client's family, emphasize follow-up, and utilize milieu-therapy principles (maximizing the group processes that help patients to learn new ways of getting on with people and abandon rationalization and denial) tend to have better results than clinics that do not. This was calculated after taking into account the differences between clinics in the types of patients that they attract. Nevertheless, the question still remains whether such ingredients might be provided just as well in day-patient and out-patient programmes.

Early intervention

While it is being decided how intensive the treatment for established alcoholics needs to be, it is fairly clear is that when early mild problem drinkers are identified, low-intensity interventions seem to be effective. Controlled studies show that counselling at health screening centres (Kristenson *et al*. 1983), in general hospital wards (Chick *et al*. 1985), and in general practice (Wallace *et al*. 1988) can influence heavy drinkers to reduce their alcohol consumption and/or their alcohol-related problems. (These three studies all used blood tests in the follow-up to confirm their results.) The alcoholic at the psychiatric clinic is only the tip of an iceberg of considerably more morbidity and mortality. We should turn more of our attention to the early detection of alcohol dependence and misuse, before severe dependence is established, social supports have been sundered, and psychiatric complications develop.

The detoxification debate

A matter of practical importance in the 'intensive versus minimal' debate is how often in-patient care is necessary to effect safe withdrawal from alcohol in the severely dependent patient. Some would claim that it is folly to give tranquillizers to out-patient alcoholics because they are likely to drink on top of the medication, and compound their problems, or perhaps take an overdose. They state that since fits and delirium can occur and these are dangerous complications, patients should be under medical care. Hayashida° *et al*. (1989) studied a group of male alcoholics of low socio-economic status who needed detoxification treatment. They excluded patients in whom a medical or psychiatric condition such as delirium tremens or a history of withdrawal fits indicated that admission was essential, and randomly allocated the remaining 164 to either in-patient or out-patient detoxification. The out-patients were evaluated medically and psychiatrically and then prescribed decreasing doses of oxazepam on the basis of daily clinic visits. The in-patient programme combined comprehensive psychiatric and medical evaluation, detoxification with oxazepam, and the initiation of rehabilitation treatment. The mean duration of treatment was attractively shorter for out-patients (6.5 days) than for in-patients (9.2 days). However, more in-patients (95 per cent) than out-patients (72 per cent) completed detoxification. There were no serious medical complications such as fits in either group. Outcome evaluation at six months revealed no differences between the groups, nor was there any difference in subsequent use of other rehabilitation services. The out-patient method cost ten times less per patient.

It is our experience too, and that of others (e.g., Stockwell *et al.* 1990), that domiciliary detoxification of alcoholics is safe, with the following guidelines:

(1) patients with a history of fits are admitted to hospital;

(2) the patient initially has to be sufficiently sober (not necessarily alcohol-free) to agree to a contract which specifies he does not drink while taking medication;

(3) generally only 24 hours' medication is supplied at one time (vitamin therapy will probably be given too);

(4) the patient is seen daily at the clinic (or at home by a community nurse if physically too unwell to travel);

(5) a large enough dose of medication is given in the first 24 hours to suppress most of the symptoms, and to permit some broken sleep (e.g., 100 mg chlordiazepoxide);

(6) further supplies of medication are only given if the patient has abstained, as confirmed ideally by a hand-held breathalyser;

(7) medication is gradually reduced to zero over five days;

(8) advice is given on avoiding drinking settings, and keeping busy (e.g., going for walks, AA meetings, Advice Centre on Alcohol).

Clinical guidelines

1. Abstinence is best, but if patients will not choose this goal or cannot achieve it, then help them set a goal of limited drinking, monitor their progress objectively, help them to find the psychological strategies to succeed, and be available if and when they change their minds and go for abstinence.

2. Assess and treat physical dependence and withdrawal, often as an out-patient.

3. Uncontroversial counselling precepts: help patients stay clear in their minds *why* they are doing something about their drinking; reinforce a sense of mastery; help them analyse how relapse could happen and practise strategies for avoiding it; guide patients towards social supports, and perhaps AA at least for several visits.

4. Offer deterrent drugs to patients in difficulties, and arrange supervision if it seems needed, e.g., at an employee's occupational health service, by the general practitioner's nurse or receptionist, by the spouse, by your clinic or community nurse. Give a dependent patient the chance to live without alcohol for a year and gain some new interests and social skills.

5. Do not miss a chance to discuss safe drinking with patients you see; exploit opportunities for early detection and counselling (there is sadly comparatively little other preventive work in psychiatry as yet).

References

Azrin, N. H. (1976). Improvements in the community reinforcement approach to alcoholism. *Behaviour Research and Therapy*, **14**, 339–48.

British Journal of Addiction (1987). Is controlled drinking possible for the person who has been severely alcohol dependent? *British Journal of Addiction*, **82**, 237–55; 841–7.

Chick, J. (1982). Do alcoholics recover? *British Medical Journal*, **285**, 3–4.

Chick, J. (1986). Treatment of alcohol dependence: abstinence or controlled drinking? *British Journal of Hospital Medicine*, **36**, 241.

Chick, J. (1990). Alcohol and drugs, treatment and control: alcohol. In *Principles and practice of forensic psychiatry* (ed. R. Bluglass and P. Bowden), pp. 929–40. Churchill Livingstone, London.

Chick, J., Lloyd, G., and Crombie, E. (1985). Counselling problem drinkers in medical wards: a controlled study. *British Medical Journal*, **290**, 965–7.

Chick, J., Ritson, B., Connaughton, J., Stewart, A., and Chick, J. A. (1988). Advice versus extended treatment for alcoholism: a controlled study. *British Journal of Addiction*, **83**, 159–70.

Christensen, J. K., Ronstead, P., and Vaag, U. H. (1984). Side-effects after disulfiram. *Acta Psychiatrica Scandinavica*, **69**, 265–73.

Cook, C. (1988*a*). The Minnesota Model in the management of drug and alcohol dependency: miracle, method or myth? Part 1. The philosophy and the programme. *British Journal of Addiction*, **83**, 625–34.

Cook, C. (1988*b*). The Minnesota Model in the management of drug and alcohol dependency: miracle, method or myth? Part 2. Evidence and conclusions. *British Journal of Addiction*, **83**, 735–48.

Costello, R. M. (1980). Alcoholism treatment and evaluation: slicing the outcome variance pie. In *Alcoholism treatment in transition* (ed. G. Edwards and M. Grant), pp. 113–27. Croom Helm, London.

Davies, D. L. (1962). Normal drinking in recovered alcohol addicts. *Quarterly Journal of Studies on Alcohol*, **23**, 94–104.

Edwards, G. and Guthrie, S. (1967). A controlled trial of in-patient and out-patient treatment of alcohol dependence. *Lancet*, **i**, 555–9.

Edwards, G., Orford, J., Egert, S., Guthrie, S., Hawker, A., Hensman, C., and Mitcheson, M. (1977). Alcoholism: a controlled study of treatment and advice. *Journal of Studies on Alcohol*, **38**, 1004–31.

Ericksen, L., Bjornstad, S., and Gotestan, K. G. (1986). Social skills training in groups for alcoholics. One year treatment outcome for groups and individuals. *Addictive Behaviours*, **11**, 309–29.

Fawcett, J., *et al.* (1987). A double blind, placebo controlled trial of lithium carbonate therapy for alcoholism. *Archives of General Psychiatry*, **44**, 248–58.

Fuller, R. K., *et al.* (1986). Disulfuram treatment of alcoholism: a Veterans Administration Cooperative Study. *Journal of the American Medical Association*, **256**, 1449–55.

Hayashida, M., *et al.* (1989). Comparative effectiveness and costs of in-patient and out-patient detoxification of patients with mild-to-moderate alcohol withdrawal syndrome. *New England Journal of Medicine*, **320**, 358–65.

Keso, L. (1988). *In-patient treatment of employed alcoholics—a randomized clinical trial on Hazelden and traditional treatment* Research Unit of Alcohol Diseases, Helsinki University Central Hospital, Helsinki.

Kristensson, H., Ohlin, H., and Hutten-Nosslin (1983). Identification and intervention in heavy drinking in middle-aged men: results and follow-up at 24–60 months of long-term study with randomized controls. *Alcoholism: Clinical and Experimental Research,* **20,** 203–9.

McCrady, B., Longabaugh, R., Fink, E., Stout, R., Beattie, M., and Ruggieri-Authelet, A. (1986). Cost effectiveness of alcoholism treatment in partial hospital versus in-patient settings after brief in-patient treatment: 12 month outcomes. *Journal of Consulting and Clinical Psychology,* **54,** 708–13.

McLachlan, J. and Stein, R. L. (1982). Evaluation of a day clinic for alcoholics. *Journal of Studies on Alcohol,* **43,** 261–72.

Miller, W. R. (1987). Behavioural alcohol treatment research advances: barriers to utilisation. *Advances in Behaviour Research and Therapy,* **9,** 145–64.

Naranjo, C. A., Sellers, E., and Lawrin, M. (1986). Modulation of ethanol intake by serotonin uptake inhibitors. *Journal of Clinical Psychiatry,* **46,** (Supplement 4), 16–22.

Nordstrom, G., and Berglund, M. (1987). A prospective study of successful long-term adjustment in alcohol dependence—social drinking versus abstinence. *Journal of Studies on Alcohol,* **48,** 95–103.

Potamianos, G., North, W. R. S., Meade, T. W., Townsend, J., and Peters, T. J. (1986). Randomized trial of community-based centre versus conventional hospital management in treatment of alcoholism. *Lancet,* ii, 797–9.

Robichaud, C., Strickland, D., Bigelow, G., and Liebson, I. (1979). Disulfiram maintenance employee alcoholism treatment: a three phase evaluation. *Behaviour Research and Therapy,* **17,** 618–21.

Rychtarik, R. G., Foy, D. W., Scott, T., Lokey, L., and Prue, D. M. (1987). Five years follow up of broad spectrum behavioral treatment for alcoholism: effects of training controlled drinking skills. *Journal of Consulting and Clinical Psychology,* **55,** 106–8.

Sanchez-Craig, M., Annis, H., Bornet, A. R., and MacDonald, K. R. (1984). Random assignment to abstinence and controlled drinking: Evaluation of a cognitive behavioral program for problem drinkers. *Journal of Consulting and Clinical Psychology,* **52,** 390–403.

Sereny, G., Sharma, V., Holt, J., and Gordis, E. (1986). Mandatory supervised Antabuse therapy in an out-patient alcoholism program: a pilot study *Alcoholism: Clinical and Experimental Research,* **10,** 290–2.

Shaw, G. K., Majumdar, S. K., Waller, S., MacGarvie, J., and Dunn, G. (1987). Tiapride in the long-term management of alcoholics of anxious or depressive temperament. *British Journal of Psychiatry,* **150,** 164–8.

Sobell, M. B. and Sobell, L. C. (1976). Second year treatment outcome of alcoholics treated by individual behaviour therapy. *Behaviour Research and Therapy,* **14,** 195–214.

Stockwell, T., Bolt, L., Milner, I., Pugh, P., and Young, I. (1990). Home detoxification for problem drinkers: acceptability to clients, relatives, general practitioners and outcome after 60 days. *British Journal of Addiction,* **85,** 61–70.

Vaillant, G. E., Clark, W., Cyrus, C., Milofsky, E. S., Kopp, J., Wulsin, V. W., and Mogielruicki, N. P. (1983). Prospective study of alcoholism treatment—an 8 year follow-up. *American Journal of Medicine*, **75**, 455–63.

Wallace, P., Culler, S., and Maines, A. (1988). Randomised controlled trial of general practitioner intervention in patients with excess alcohol consumption. *British Medical Journal*, **297**, 663–8.

15

Childhood sexual abuse: how can women be helped to overcome its long-term effects?

MIKE HOBBS

An increasing number of women seek psychiatric treatment for the lasting destructive effects of sexual abuse in childhood. How can these women be helped? While there is much clinical experience, and a limited amount of experimental data to guide us, the clinician is inevitably faced with a number of dilemmas.

Some dilemmas

1. Despite increased public awareness of childhood sexual abuse (CSA), and some reduction in social prejudice towards its victims, many women (and perhaps even more men) are still unable to seek help openly for its lasting psychological effects.

2. Some victims of CSA, having been referred for treatment of a particular psychological disorder, disclose their childhood experience of sexual abuse in the course of treatment. Others seek and undergo treatment without mention of this history.

3. For a significant number of victims, the inability to disclose their experience is a product of having put all memory of the sexual abuse out of mind: that is, they have *repressed* memory of it.

4. The adult victims of CSA present a broad range of psychopathology, both in terms of its features and its severity. Despite this, there is still a tendency among psychiatrists to view these women as a *unitary population* for whom there is a single universal treatment.

5. When aware of a history of CSA, many clinicians are uncertain whether to address that past experience or to focus exclusively on the current psychological disorder.

6. Working with the adult victims of CSA is a complex and stressful endeavour which generates a wealth of disturbing feelings in the clinician. These reactions may inhibit or destroy the treatment exercise.

The controversy

Let me state my central proposition. Women who have been sexually abused in childhood *can* be helped to overcome its long-term effects; but, because there is no unitary psychopathology which results from childhood sexual abuse (CSA), there can be no universal treatment. Treatment must go beyond the manifest symptoms or syndrome with which the woman presents in order to address directly the aetiological experience of sexual abuse itself. Clinicians need to monitor their own emotional reactions to the patient's story and behaviour.

Epidemiology and effects

Prevalence

Studies of the prevalence of CSA in *community* samples of adult American women have shown rates of 6–62 per cent (Finkelhor 1986). In Great Britain, Baker and Duncan (1985) found that 12 per cent of a commumity sample of women had been sexually abused in childhood, and another 12 per cent declined to answer the question. Forty per cent of the reported cases had been subject to physical sexual contact.

Higher prevalence rates have been found in *clinical* samples. The rates for men are about half of those for women.

Lasting effects

CSA is associated with a range of immediate and lasting psychiatric effects. A significant proportion of victims (perhaps 20 per cent) experience persisting severe psychological problems. In view of the large size of the population under consideration, the implications for the mental health of the female population are enormous.

Compared with control populations, adult women who were abused sexually as children are more likely to experience and manifest depression, anxiety, feelings of isolation and stigmatization (particularly guilt and shame), poor self-esteem, dissociation, self-destructive behaviours, and substance abuse. Problems in interpersonal relationships are evidenced by pervasive mistrust, sexual maladjustment, and a tendency towards revictimization. Within relationships, a broad range of psychosexual difficulties have been reported including aversion, dissatisfaction, impaired motivation, inhibited arousal, and anorgasmia (Finkelhor 1986; Fromuth 1986; Briere 1988; Wyatt and Powell 1988; Jehu 1989).

Specific abuse: effect characteristics

Recent studies have demonstrated that *prolonged* experience of CSA, an *early age* at onset and a *late age* at cessation, abuse by *father* rather than a non-relative, *physical* sexual contact (compared with exhibitionistic, voyeuristic, or verbally suggestive acts), and the use of *force* by the perpetrator are associated with later psychological disorder.

Briere (1988) showed the particular pathogenic impact of abuse which involved physical *violence*, bizarre or *repugnant acts,* or *multiple perpetrators*. He demonstrated also that *intercourse* during abuse predisposes to particularly high levels of *dissociation* and of *suicidal* ideation and action. Presumably other forms of penetration, including anal and oral variants, have similar or worse effects. Briere hypothesized that all sexual abuse is 'traumagenic' but that its traumatic impact is heightened by these specific characteristics.

Kroll (1988) has suggested that childhood sexual abuse is a major aetiological factor in the development of borderline personality disorder, and has pointed to the particular relationship between a history of sexual abuse and the brief psychotic episodes which characterize the more severe forms of borderline personality functioning. This and other features of borderline personality functioning, such as dissociation, self-destructive actions and habits, severe interpersonal maladjustment, and identity disturbance are understandable in terms of the damage to the child's developing self which is inflicted by sexual abuse, particularly when it occurs very early in life and is accompanied by penetration of the child's body, physical violence, and the boundary violations inherent in incest.

Problems of disclosure

Betrayal and 'accommodation'

I have suggested already that many women who were sexually abused in childhood do not disclose this fact when seeking help for its long-term effects. The adult victim's difficulty in disclosing her childhood experience repeats the dilemma of the child who, already traumatized by the sexual abuse itself and by the secrecy enforced by the perpetrator, may be traumatized further by the disbelief, blame, and rejection of the adults to whom she turns for help, perhaps her mother. Her experience of fear and pain is compounded by a sense of betrayal by the very adults upon whom she relies most.

This betrayal, as much as the abuse itself, disconfirms and damages the child's emerging self and so contributes to later disorders of self-concept and personality. The child's helplessness and isolation in the face of the threats of the abuser and the betrayal of others leads her to 'accommodate' (Summit 1983) the abuse. She cannot disclose her experience or escape from it. She may even deny it. Repression, denial, and dissociation are common ways of attempting unconsciously to deal with this trauma.

Further traumatization in the treatment context

It is not just the child who accommodates in this manner. Many victims of CSA have been rejected by the professionals to whom they have turned for help, sometimes many years later. The commonest problem encountered is scepticism or disbelief. Two recent examples in my own clinical practice serve to demonstrate this. One young woman had told her general practitioner of the sexual abuse to which she had been subjected by her father for many years, and which continued forcibly even after she fled from home. The general practitioner, her family doctor, simply refused to believe this of her father who was a respected professional man and church elder in their community. Instead he accused the patient of immaturity and malice. A second example was demonstrated by a referral letter which I received which began 'this 36-year-old woman *alleges* that her stepfather abused her sexually in childhood', as if her experience needed to be proven before we could respect and respond to the patient's claims.

Our refusal straightforwardly to believe the abused individual is a further traumatization, compounding the primary betrayal of her need to trust the adults upon whom she relied in childhood. It damages further her self-image, capacity for trust, and sense of self-determination.

The most extreme example of further traumatization is for the victim to be violated again sexually by the professional person to whom she turns, by innuendo, voyeuristic exploitation, or even sexual assault. There is evidence to suggest that sexual exploitation occurs all too commonly even in the therapy setting (Kardener *et al.* 1973). Such abuse has been described as a product of the therapist's unconscious identification with the abuser, a dynamic which is common and problematic especially for male therapists working with female victims (Frosh 1987).

Of course some women *may* have fabricated their stories of childhood sexual victimization; and, of course, some women who *were* sexually abused in childhood do behave in a sexually flirtatious and seductive manner, even with their therapists. Such behaviours remind us powerfully of those experiences which led Freud to abandon his 'seduction theory' for his theories of infantile sexuality (Masson 1985; Skues 1987). Nevertheless it is the responsibility of the professional person to know enough about the effects of CSA and about unconscious mental processes, to be aware of these dynamics, and not to compound the patient's problems by an unprofessional manner or conduct.

Disguised presentation

The woman's own repression of her traumatic experience or her fear of disbelief in those to whom she might turn mean that the lasting effects of sexual abuse are often presented 'in disguise' (Gelinas 1983). The clinician is faced with a diagnostic problem, for failure to recognize the antecedents of the presenting disorder will undermine the likely effectiveness of treatment. Gelinas identified the most common disguised presentation as a characterological depression complicated by dissociative features, impulsive and self-destructive behaviours. She suggests that a history of 'parentification' is a common clue to the history of sexual abuse, evidenced by premature and heavy housekeeping or child care responsibilities during childhood and adolescence.

Treatment

Principles of treatment

1. *Respect and belief* The first principle of treatment for a woman who was sexually abused in childhood is that her story should be believed by her therapist, and that she should be able to expect freedom from further abuse or exploitation in therapy.

2. *Assessment and selection of treatment* A thorough assessment of the woman's problems and needs is essential, including reference to the trauma itself (though the details of her experience may be disclosed only later in therapy, if at all); its early and later effects on the individual and her social network; her experience of further victimization in childhood or adulthood; her current interpersonal functioning, including her sexual functioning and orientation; her personality structure, including character-ological defences; and the availability to her of current supportive and confiding relationships. It is upon the information obtained from detailed assessment that the goals and choice of treatment will be determined.

3. *Protection* A third principle of management, essential when treating children but also when treating adults, is the need for a decisive protective intervention if the sexual abuse continues. This might involve other members of the family, including the perpetrator. Protective intervention is necessary particularly when sexual abuse which continues into adulthood is characterized by violence and sadism which leaves the victim frightened and helpless.

4. *Strategy of intervention* Wheeler and Berliner (1988) contrast 'approach' and 'avoidant' strategies for coping with stress and anxiety. They suggest that treatment interventions for the lasting effects of CSA will generally need to be 'approach' strategies which are directed towards the active, but gradual, tackling of anxiety and traumatic experience and which minimize counterproductive denial and avoidance. However they recognize explicitly that avoidance strategies may be therapeutic in certain situations in order to prevent the individual from feeling overwhelmed by anxiety generated by 're-living' her experience.

5. *Pacing* Related to the above is the need to pace the therapy according to the patient's needs and coping resources. The woman, in effect, is given 'permission' to speak about her experience of being abused, but at her own

pace. Pressure on the patient for self-disclosure or confrontation of mal-adaptive beliefs and behaviours can be traumatizing.

6. *Preventive emphasis*   It is not enough in treatment to expose and to defuse of their traumatic potential the memories of sexual abuse. It is necessary to overcome 'the past in the present', that is, the woman's inner compulsion to repeat her experience of victimization through, at best, her continued denigration of herself and, at worst, self-harm, promiscuity, prostitution, or vulnerability to further rape or assault. Fundamental to the preventive emphasis is the need to enable the woman to recognize and avoid partners who are likely to abuse her or her/their own children in a manner which repeats her previous experience; and this preventive work also necessitates that the woman recognize in herself her own propensity to attract or promote abusive behaviour in the men with whom she relates. The transgenerational cycle of sexual abuse has been well described by Gelinas (1983).

7. *De-sexualization of the abusive experience*   Faria and Belohlavec (1984) suggest that, as a fundamental principle of treatment, the victim's experi-ence should be construed in terms of a power relationship rather than sexuality. Some women do report having experienced sexual feelings and even pleasure when they were abused; but the experience was still the product of an older person's exploitation of his power over her and a betrayal of his responsibility towards her.

8. *Therapist's emotional outlets*   Working with CSA victims is stressful, and generates powerful counter-transference reactions in the therapist (Ganzarain and Buchele 1988). The therapist requires outlets for his or her emotional reactions in order to safeguard the patient and the therapy (Faria and Belohlavek 1984). Even with support, therapists may be unable to sustain close work with victims of CSA indefinitely without suffering a 'pervasive and leaden sense of inner desolation', an emotional response 'which mirrors faithfully the inner desolation so well known to these patients' (MacCarthy 1988).

Overview of treatment

Women who are suffering the lasting effects of CSA have been treated by clinicians from a variety of professional and therapeutic backgrounds, and in a variety of settings. Published reports indicate that some common treat-ment principles apply even when the approaches are apparently disparate. There is no evidence to suggest that one form of treatment is more effective than another. In view of the wide range of experiences, symptomatology, and severity of problems which are manifest by this population of women, I believe that there will always be a need for a range of treatments.

Women who were abused seem to benefit from talking of their experience and from the associated catharsis, but only if they are provided with a therapeutic relationship and a strategy for coping with the overwhelming memories and feelings which are recovered. For this reason, simple reflective counselling can prove harmful.

The most numerous reports of therapeutic intervention refer to time-limited group treatments (Tsai and Wagner 1978; Herman and Schatzow 1984; Cole 1985; Wildsmith and Wolfers 1987), some of which have been conducted in conjunction with other treatment approaches, such as individual therapy (Goodman and Nowak-Scibelli 1985) and family therapy (Deighton and McPeek 1985). Two centres have reported experience of working with the adult victims of CSA in long-term group psychotherapy (Blake-White and Kline 1985; Ganzarain and Buchele 1986, 1987, 1988).

Many CSA victims have been treated in analytic psychotherapy. This work has illuminated some of the dynamic processes involved in the treatment of the victims of sexual abuse which might apply both in the individual and group treatment settings (Frosh 1987; MacCarthy 1988).

Recently a number of publications have described a cognitive-behavioural approach to understanding and treating the long-term effects of sexual abuse (McCarthy 1986; Jehu *et al.* 1984, 1986; Jehu 1988, 1989), conducted generally on a one-to-one basis but sometimes including partners or involving assertiveness training in a group format.

Therapeutic relationship and goals of treatment

Faria and Belohlavek (1984) have identified some of the fundamental issues for therapy which apply regardless of the treatment modality adopted. They suggest that every therapist should proceed from a particular frame of reference and their own incorporates some of the principles identified above. In common with other practitioners, they emphasize the importance of establishing a sound therapeutic relationship and recognize that this may be difficult in view of the pervasive mistrust and interpersonal awkwardness of some victims. The goals of therapy include the identification and tackling of avoidant patterns in relationships, promotion of the patient's self-esteem, her constructive expression of anger, the development of strategies for her gaining control over self-destructive and self-defeating behaviours, the development of mutually supportive relationships, and the promotion of more positive body image and sexual response.

Jehu (1989) reminds us that, although the patient's problems originated in her earlier experience of childhood sexual abuse, 'It is the contemporary conditions that initiate and maintain the problems that require therapeutic attention'. All of the treatments advocated for this patient population do indeed focus upon current problems, even though reference may be made to past experience in order to understand and tackle them more effectively.

Cognitive-behavioural treatment

Jehu (1988), in the most detailed available account, divides the components of treatment into 'certain general conditions', including the therapist–client relationship, and more specific procedures for tackling particular aspects of the patient's difficulties. He emphasizes the importance of the *initial assessment* of the patient's experience and current problems, in order to identify the objectives and techniques of treatment successfully. The *formulation* of the patient's problems includes:

(1) precise specification of the nature of the problems;

(2) hypotheses about the current events which evoke and maintain them;

(3) historical or developmental factors that have predisposed the patient to react to her abuse in these ways: and

(4) appraisal of the resources available to her.

The *goals of treatment* are identified and agreed mutually by therapist and patient, thus maximizing the latter's collaboration in the treatment process.

Jehu argues for a treatment which is *individualized* for each patient according to her current problems, her personal characteristics, and her present life situation. Thus, in addition to working on problems associated with mood and self-image, he might advocate involvement in the therapy of the patient's current partner and/or her participation in an assertiveness training group. The various components prescribed are *sequenced* in a graded manner so as to maximize the patient's achievement and experience of success at each stage. He stresses the importance of the *patient's motivation* in maintaining the therapeutic relationship and ensuring a positive outcome; and he identifies some ways by which the patient's motivation can be re-harnessed when resistance are encountered during treatment.

The respective advantages and disadvantages of female and male therapists are examined by Jehu. He suggests that, even if the patient's mistrust of men leads her initially to prefer a female therapist, there can be an advantage in the patient working later with a male therapist, either singly or in a female/male co-therapy pair. Her work with a non-exploiting male therapist can offer a powerful corrective experience.

Jehu's cognitive-behavioural model of treatment is based on the division of the patient's current problems into three categories, each of which can be construed in terms of cognitive and behavioural theories—*mood disturbance, interpersonal problems*, and *sexual dysfunction*. Specific techniques are selected for the treatment of each category of problem.

Mood disturbance

On the basis that mood disturbances are underpinned by the woman's distorted beliefs about herself and her world, formed during or as a result of her childhood experience of sexual abuse, these are treated by a process of cognitive restructuring. The rationale for treatment is explained, the distorted beliefs are identified, and then more accurate and realistic alternative beliefs are elicited, evaluated, and rehearsed in the treatment setting. Jehu identifies the most common cognitive distortions as the self-blame and self-denigration with which are associated feelings of guilt, shame, worthlessness, and self-disgust.

Interpersonal problems

The interpersonal problems identified by Jehu revolve around the common themes of isolation, insecurity, discord, and inadequacy. The treatment procedures employed are based on a psycho-educational and behavioural strategy of communication skills, problem-solving, anger control, and assertiveness training.

Sexual dysfunction

Not surprisingly a wide variety of sexual dysfunctions have been recorded, and Jehu explores their causes in relation to a number of factors including the mood disturbances and interpersonal problems which are the other characteristic long-term sequelae of CSA. He advocates a range of techniques for tackling specific aspects of these dysfunctions, based on behavioural strategies (relaxation training, thought and image stopping, systematic desensitization, etc.) and cognitive restructuring.

Results of treatment

In his report of the treatment of 36 of the 51 women who entered his study, Jehu records some impressive results in relation to mood disturbance. The mean duration of therapy was 21.2 weeks (range 3–47 weeks), and the mean duration of follow-up was 57.6 weeks (range 8–135 weeks). Significant levels of mood-related distorted beliefs were recorded at assessment by 94 per cent of cases, but by only 13 per cent at the end of treatment and 5 per cent at final follow-up. At assessment 58 per cent of victims showed clinically significant scores on the Beck Depression Inventory, and these were reduced to 8 per cent at termination and 5 per cent at final follow-up. On small numbers ($n = 15$) Jehu demonstrated clinically significant impairment of self-esteem in 86 per cent of cases at assessment, which was reduced to 53 per cent at termination and 40 per cent at final follow-up. Among respondents to a consumer satisfaction survey ($n = 41$), more than 90 per cent of patients were partially or completely satisfied by the results of their treatment.

Group therapies

The rationale for group treatments of women who were sexually abused in childhood is not primarily one of economic expediency. Much more importantly, a treatment group (whether short-term or long-term, structured or unstructured) affords to its members direct opportunities for overcoming their isolation and stigmatization and for generating trust and openness in relationships. As Blake-White and Kline (1985) have observed, 'For the first time [the woman's] incestuous experience makes her one of a group, rather than the deviant. . .'. The well-recognized group therapeutic factors of acceptance, universality, catharsis, interpersonal learning, and self-understanding (Bloch and Crouch 1985) operate powerfully in groups composed homogeneously of women who were abused in childhood. Different aspects and combinations of these therapeutic processes are promoted in the various styles of group work undertaken. However it has been found repeatedly, whether by impression or by more objective evaluation (Tsai and Wagner 1978; Herman and Schatzow 1984), that women rate the opportunity to meet with and share their feelings and experiences with other victims as the most helpful and therapeutic aspects of group treatment.

Time-limited group therapy

There is a convincing rationale for time-limited group therapy, though several authors recognize that short treatments only initiate the process of change. Short-term treatments minimize the opportunities for regression, emphasize the patient's responsibility and coping, and highlight issues associated with separation and loss, all of which are valuable in approaching the treatment of this patient population. In addition the time-limited model of group treatment promotes *active* focus on the women's experience of abuse and the combat of their previous patterns of avoidance and accommodation; and the model also permits a heightened awareness of boundary issues (most particularly time, of course) which helps to reverse some of the boundary disturbance implicit in the violations of sexual abuse.

Time-limited group therapy combats the passivity which is so often a defence against chronic feelings of helplessness and powerlessness. Each woman's distorted views of herself are exposed, examined, and remedied through interaction with the other members of the group. Her prolonged experience of secrecy, isolation, and powerlessness can be reversed in a way which, very often, enables her to develop a new-found sense of belonging. This gives rise to appropriate regret and grieving when the group ends.

Results of time-limited group therapy

The time limit and structure of short-term groups varies. Tsai and Wagner (1978) reported a four-session model in which 50 women had been treated in ten groups by female/male co-therapy pairs. At the end of the group, each woman was asked to rate on a seven-point scale (amongst other factors) the overall helpfulness of the group experience (mean score 6.0), diminution in their feelings of guilt (mean 6.1), enhanced self-acceptance (mean 5.8), and improvements in their relationships with current partners (mean 5.2).

Herman and Schatzow (1984) reported the treatment of 30 women in groups which met weekly for ten sessions. Twenty-eight completed therapy, and 20 responded to a postal follow-up survey in which 85 per cent reported improvements in self-esteem, 80 per cent indicated that they felt less ashamed and guilty, and 75 per cent that they were less isolated. Lesser improvements were noted in relationships, sexual enjoyment, and work performance. The authors concluded that this treatment model is particularly effective in resolving problems associated with shame, secrecy, and stigmatization.

Long-term group therapy

The advocates of longer-term group therapy point to the stunted personality development and severe psychopathology of a significant proportion of women who were abused, perhaps particularly when the abuse involved those features which have been demonstrated to have high impact. Certainly it seems unlikely that the more severe manifestations of dissociation, self-destructive actions, and interpersonal difficulties will be modified significantly in any short-term treatment approach; and the opportunities within the brief treatment format for personality development and change are very limited.

Blake-White and Kline (1985) reported a long-term, open-ended group in which dissociative processes were tackled in a way that gradually permitted each woman to own and integrate aspects of herself which were first identified in other members within the group. In a superficial analysis of 54 patients, 37 per cent were reported to have dropped out of the group in five sessions or less; another 33 per cent terminated their attendance in 20 sessions or less because they felt that they had benefited from the group as much as they needed, though the therapists did not feel that their treatment had been concluded satisfactorily. The other 30 per cent had attended the group for more than 20 sessions, and 13 per cent of them for more than 50. The authors implied that these women demonstrated lasting beneficial changes in their relationships, life-styles, and personalities.

Ganzarain and Buchele (1988) reported their experience of working with 25 patients in a long-term analytic therapy group, membership of which required a commitment to attend weekly sessions for at least one year. Each of the patients was in concurrent individual analytic psychotherapy with other therapists. Unlike some of the other studies reported, these authors selected for their group patients who demonstrated the more severe manifestations of psychopathology, including borderline personality functioning. Many of the patients displayed a very limited ability to put their experiences and concerns into words so that, as treatment progressed, acting-out was frequent and dramatic. No empirical data were offered to demonstrate the effectiveness of treatment, but the authors described very clearly the difficulties and principles inherent in treating the most severely damaged victims of sexual abuse.

Conclusion

Both from reviewing the literature and from personal experience, my impression is that each of the models outlined has a place in the initial treatment of women who suffer the lasting effects of childhood sexual abuse. Each treatment approach addresses the central themes of victimization through its own philosophy and techniques, though different facets are emphasized by each.

We need to know more about the respective indications for each of the specific treatments, as well as for their combinations. We need also to consider when it would be advantageous to include the woman's partner in treatment. It is evident that short but active treatments, whether conducted in a group format or in individual treatments involving cognitive-behavioural techniques, can effect significant and lasting improvements for some women. It is my belief, however, that substantial modifications of personality are necessary in order to exert any lasting beneficial impact upon the most severe late manifestations of CSA, in particular when these are associated with substantial dissociation and fragmentation of the personality. Such ambitious change can be achieved only through longer-term analytic psychotherapy in a group or individual setting, or by a combination of both.

Clinical guidelines

1. Women who were sexually abused in childhood can be helped to overcome its lasting destructive effects.

2. To be effective, treatment must address the experience of sexual abuse as well as the symptoms and psychopathology which it generated; and it must provide a strategy for coping with the memories and feelings which are uncovered.

3. Short-term treatments are effective in ameliorating a wide range of symptoms, but cannot be expected to eradicate the deeper layers of shame or to promote substantial psychological development in those women whose personalities were stunted or distorted by their experience of abuse.

4. Cognitive-behavioural methods are effective in the treatment of mood disorders, psychosexual dysfunction, and some interpersonal problems resulting from CSA.

5. Time-limited group therapy is effective in overcoming the woman's sense of powerlessness, isolation, and guilt; and can produce improvement in self-image and in existing relationships.

6. Longer-term analytic group therapy or analytic psychotherapy is necessary for the treatment of the more severe dissociative and self-destructive features, and of severe interpersonal maladjustments and personality disturbance which follow CSA.

7. Concurrent individual and group analytic therapies are effective and perhaps synergistic.

References

Baker, A. W. and Duncan, S. P. (1985). Child sexual abuse: a study of prevalence in Great Britain. *Child Abuse and Neglect*, **9**, 457–67.

Blake-White, J. and Kline, C. M. (1985). Treating the dissociative process in adult victims of childhood incest. *Social Casework: the Journal of Contemporary Social Work*, **66**, 394–402.

Bloch, S. and Crouch, E. (1985). *Therapeutic Factors in Group Psychotherapy*. Oxford University Press.

Briere, J. (1988). The long-term clinical correlates of childhood sexual victimisation. *Annals of the New York Academy of Sciences*, **528**, 327–34.

Cole, C. L. (1985). A group design for adult female survivors of childhood incest. *Women and Therapy*, **4**, 71–82.

Deighton, J. and McPeek, P. (1985). Group treatment: adult victims of childhood sexual abuse. *Social Casework: the Journal of Contemporary Social Work*, **66**, 403–10.

Faria, G. and Belohlavek, N. (1984). Treating female adult survivors of childhood incest. *Social Casework: the Journal of Contemporary Social Work*, **65**, 465–71.

Finkelhor, D. (ed.) (1986). *A sourcebook on child sexual abuse*. Sage, Beverly Hills, CA.

Fromuth, M. E. (1986). The relationship of childhood sexual abuse with later psychological and sexual adjustment in a sample of college women. *Child Abuse and Neglect*, **10**, 5–15.

Frosh, S. (1987). Issues for men working with sexually abused children. *British Journal of Psychotherapy*, **3**, 332–9.

Ganzarain, R. and Buchele, B. (1986). Countertransference when incest is the problem. *International Journal of Group Psychotherapy*, **36**, 549–66.

Ganzarain, R. and Buchele, B. (1987). Acting out during group therapy for incest. *International Journal of Group Psychotherapy*, **37**, 185–200.

Ganzarain, R. and Buchele, B. (1988). *Fugitives of incest: a perspective from psychoanalysis and groups*. International Universities Press, Madison, Connecticut.

Gelinas, D. (1983). The persisting negative effects of incest. *Psychiatry*, **46**, 312–32.

Goodman, B. and Nowak-Scibelli, D. (1985). Group treatment for women incestuously abused as children. *International Journal of Group Psychotherapy*, **35**, 531–44.

Herman, J. and Schatzow, E. (1984). Time-limited group therapy for women with a history of incest. *International Journal of Group Psychotherapy,* **34,** 605–16.

Jehu, D. (1988). *Beyond sexual abuse: therapy with women who were childhood victims.* Wiley, Chichester.

Jehu, D. (1989). Sexual dysfunctions among women clients who were sexually abused in·childhood. *Behavioural Psychotherapy,* **17,** 53–70.

Jehu, D., Gazan, M., and Klassen, C. (1984). Common therapeutic targets among women who were sexually abused in childhood. *Journal of Social Work and Human Sexuality,* **3,** 25–45.

Jehu, D., Klassen, C., and Gazan, M. (1986). Cognitive restructuring of distorted beliefs associated with childhood sexual abuse. *Journal of Social Work and Human Sexuality,* **4,** 49–69.

Kardener, S., Fuller, M., and Mensh, I. (1973). A survey of physician's attitudes and practices regarding erotic and non-erotic contact with patients. *American Journal of Psychiatry,* **130,** 1077–81.

Kroll, J. (1988). *The challenge of the borderline patient: competency in diagnosis and treatment.* Norton, New York.

McCarthy, B. W. (1986). A cognitive-behavioural approach to understanding and treating sexual trauma. *Journal of Sex and Marital Therapy,* **12,** 322–39.

MacCarthy, B. (1988). Are incest victims hated? *Psychoanalytic Psychotherapy,* **3,** 113–20.

Masson, J. M. (1985). *The assault on truth: Freud's suppression of the seduction theory.* Penguin, Harmondsworth.

Skues, R. (1987). Jeffrey Masson and the assault on Freud. *British Journal of Psychotherapy,* **3,** 305–14.

Summit, R. (1983). The child sexual abuse accommodation syndrome. *Child Abuse and Neglect,* **7,** 177–93.

Tsai, M. and Wagner, N. (1978). Therapy groups for women sexually molested as children. *Archives of Sexual Behavior,* **7,** 417–27.

Wheeler, J. R. and Berliner, L. (1988). Treating the effects of sexual abuse on children. In *Lasting effects of child sexual abuse* (ed. G. E. Wyatt and G. J. Powell), pp. 227–47. Sage, Newbury Park.

Wildsmith, S. and Wolfers, O. (1987). Repairing the damage. *Community Care,* **672,** 24–5.

Wyatt, G. E. and Powell, G. J. (eds) (1988). *Lasting effects of child sexual abuse.* Sage, Newbury Park.

16

Rape: can victims be helped by cognitive behaviour therapy?

EDNA B. FOA and BARBARA OLASOV ROTHBAUM

Rape is a traumatic event which is often emotionally debilitating. The physical, cognitive, and behavioural sequelae of rape are characteristic of post-traumatic stress disorder (PTSD) (American Psychiatric Association 1987). The DSM-III-R criteria include: fear and avoidance of rape-related situations, disburbance in sleep patterns, nightmares, exaggerated startle responses, intrusive unpleasant imagery, impairment in concentration or memory, and guilt. For many victims, PTSD symptoms decline within three months, although the course of these responses and their pattern of decline vary between individuals. A relatively large proportion of victims, however, continue to exhibit symptoms which disrupt daily functioning. With the high incidence of rape and the consequent high prevalence of rape-related chronic PTSD sufferers, the discovery of effective treatment strategies for these victims is of major importance.

Crisis intervention and group psychotherapy are the most common procedures used in rape crisis centres. However, research on treatment outcome for rape victims has focused almost exclusively on cognitive-behavioural methods. The treatments employed have concentrated primarily on assault-related symptoms as targets for the treatment interventions. Such targets have included fear or anxiety and accompanying avoidance patterns, depression, and social and sexual functioning. Some investigations have tested a specific intervention, examining the effects of one procedure (e.g., systematic desensitization), whereas others have developed and studied behavioural treatment packages (e.g., stress inoculation training).

Systematic desensitization

In a series of nine cases, Turner (1979) found that systematic desensitization (SD) reduced fear, anxiety, and depression. Frank and Stewart (1983, 1984) corroborated these findings with a sample of 17 assault victims. Their procedure deviated from the standard use of SD in several ways. First, the scenes were composed of long narratives rather than the customary short scenes. Second, the scenes included pleasant descriptions (e.g., 'It's a bright spring day, the trees are budding, and the sun is warm' (p. 255,

1983)). Third, the therapist moved from one hierarchy to another in the same session. The usual practice is to complete one hierarchy before moving to the next. Fourth, whereas usually each item is presented repeatedly until it fails to evoke anxiety and only then is the next item presented, Frank and Stewart presented each scene only twice.

Fourteen sessions of SD conducted as described above decreased the targeted fear. The authors noted that 75 per cent of their subjects voluntarily exposed themselves *in vivo* to situations previously desensitized in imagination. These results are quite impressive. However, in the absence of a control group, it is difficult to interpret them. Since some clients received treatment immediately following the rape, then improvement may have reflected the natural course of symptom reduction over the first few months following assault.

Cognitive therapy

The effects of cognitive therapy (CT) targeted at depression and anxiety were studied in 25 rape victims who entered treatment an average of two weeks after their assault (Frank and Stewart 1984). This treatment, based on Beck's procedure, included self-monitoring of activities (mastery and pleasure responses), graded task assignments (e.g., going out alone), and modification of maladaptive cognitions. Cognitive techniques were included to help the client identify distorted beliefs and test their reality. The outcome of this therapy was similar to that of SD: anxiety and depression decreased significantly.

In a recent article (Frank *et al.* 1988), data from 84 subjects, some of whom had participated in previous studies, were reported. Subjects were randomized into two groups: CT and SD. In each treatment modality, some subjects were treated soon after the rape (mean of 20 days) while others were treated several months post-assault (mean of 129 days). No differences between SD and CT were detected, nor did the time elapsed from rape influence treatment outcome. Although the absence of a control group makes interpreting these results problematic, the improvement evidenced by delayed treatment seekers (who were more symptomatic than the immediate treatment seekers) suggests that both treatments had active therapeutic ingredients.

Flooding

Treatment by flooding (prolonged imaginal or *in vivo* exposure to disturbing fear cues) has been successful in alleviating anxiety disorders including PTSD in Vietnam veterans (Keane *et al.* 1989). Few reports exist about its application to rape victims. Imaginal flooding proved effective for a series of four physical and sexual victims and for one incest victim. Despite these successes, the use of this procedure in sexual assault victims has drawn pointed criticism (Kilpatrick *et al.* 1982). The following concerns have been expressed about flooding:

(1) it focuses too much on anxiety as a target for change to the exclusion of irrational cognitions;

(2) it may result in an inappropriate reduction of anxiety to non-consensual forced sex;

(3) it may result in higher treatment drop-out rates due to the aversiveness of the procedure; and

(4) it fails to address the development of coping strategies.

In a rebuttal, Rychtarik *et al.* (1984) noted that irrational thoughts, anxiety, and avoidance are not independent of each other. Therefore, reduction of anxiety to rape-related cues may result in the amelioration of associated negative irrational cognitions. Moreover, as noted by Foa and Kozak (1986), cognitive mechanisms underlie some of the changes produced by exposure (flooding) and therefore are expected to influence fear, avoidance, and associated cognitions. With respect to the second concern, flooding has often been directed at unduly intense fears of realistic concerns (e.g., death, falling from heights). Decreasing such fears does not, however, lead to carelessness about one's safety. In the same vein, decreasing a woman's emotional distress to the memory of her rape need not lead to disregard for her well-being. As to the third point, we have not found that flooding resulted in higher drop-out rates than less stress-producing therapies. Finally, Rychtarik and colleagues pointed out that the experience of symptom reduction via flooding may constitute by itself a coping strategy which can be utilized systematically by clients when experiencing intense fear.

Stress inoculation training

Stress inoculation training (SIT) was developed for victims who remained highly fearful three months after being raped (Kilpatrick *et al.* 1982). The original programme, which included 20 therapy hours and homework assignments, consisted of education and training in coping skills. It began with a two-hour educational session in which the treatment programme was described and the rationale and the theoretical basis for the treatment were discussed. The programme was presented as a cognitive-behavioural approach to the management of rape-related fear and anxiety which utilized coping skills to reduce anxiety. Rape-related fear was explained as a classical conditioning phenomenon. Anxiety was described as a multi-channel system which included behavioural/motor, cognitive, and physiological responses.

The acquisition and application of the coping skills phase began with

Jacobsonian deep muscle relaxation training and breathing control. The relaxation sessions were taped for home practice and continued until the client was capable of relaxing herself within a short period of time in a variety of situations. The breathing control exercises emphasized diaphragmatic breathing similar to the exercises taught in yoga or Lamaze natural child-birth classes. Next, the client was taught communication skills through role-playing. The therapist first played the client's role in order to model appropriate behaviour. Roles were then reversed, thus giving the client an opportunity to practise the new behaviour. Covert modelling was also taught. This technique was similar to role-playing but used imagery rather than *in vivo* practice. Since the client was asked to imagine completing problematic situations successfully, training in this procedure began only after the client had successfully applied the previously learned skills.

To control obsessive thinking, thought-stopping was taught. First, the client was asked to generate the troublesome thoughts for 30–45 seconds. The therapist then shouted 'Stop!' This process was repeated several times. Next the client was asked to shout aloud 'Stop!' after the thought was generated. Finally, the client was asked to produce the thought, then silently verbalize the word 'Stop!'

The last technique, guided self-dialogue, was considered the most important. The therapist taught the client to focus her internal dialogue and to identify irrational, faulty, or negative self-statements. Rational and positive statements were generated and substituted for the negative ones following Meichenbaum's stress inoculation training.

In an empirical investigation of this programme, Veronen and Kilpatrick (1982) treated 15 rape victims, who showed elevated fear and avoidance to specific phobic stimuli three months post-rape, with the 20-hour version of SIT. A clear treatment effect emerged, with improvement on rape-related anxiety, general anxiety, tension, and depression.

The SIT programme has been modified by Kilpatrick and his colleagues. In particular, it was shortened from 20 to eight therapy hours and the application of coping skills was broadened to include problems not directly related to the rape.

Comparison between treatments

Thirty-seven rape victims participated in a study by Resick *et al.* (1988) comparing six two-hour sessions of three types of group therapy with a naturally occurring wait-list control group. The treatment conditions were SIT, assertion training, and supportive psychotherapy plus information. SIT was similar to that described above with two exceptions: (1) cognitive restructuring, assertiveness training, and role-play were excluded since they were used in the assertion training, and (2) exposure *in vivo* was added.

Assertion training began with an educational phase which included an explanation of how assertion can be used to counter fear and avoidance. Issues concerning social support were discussed within the context of assertiveness. The specific techniques were adopted from Lange and Jakubowski (1976) and from rational emotive therapy (Ellis 1977). Training included behavioural rehearsal via role-play with feedback regarding performance. The third therapy, supportive psychotherapy plus information, consisted of an educational phase after which participants selected topics for discussion. Topics included the reactions of others to their assault, the degree of support they encountered, as well as assault-induced anxiety. The group was designed to validate the members' reactions to sexual assault as well as to offer general support. No behavioural techniques were employed.

All three treatments were found effective in reducing symptoms, with no group differences evident. Improvement was maintained at six-month follow-up on rape-related fear measures, but not on depression, self-esteem, and social fears. No improvements were found in the wait-list control group.

A study comparing SIT, exposure treatment, supportive counselling, and a no-treatment control is being conducted by Foa and Rothbaum. All clients are at least three months post-assault and meet diagnostic criteria for PTSD. SIT treatment is conducted in a similar manner to that described by Kilpatrick *et al.* (1982). It consists of nine 90-minute sessions delivered twice weekly. The treatment programme is comprised of information gathering, education and treatment planning, brief breathing retraining, deep muscle relaxation, thought-stopping, cognitive restructuring, guided self-dialogue, covert modelling, and role-play. This programme deviates from the manner in which it is currently being conducted by Kilpatrick and his colleagues in that instructions for *in vivo* exposure to feared situations are not included.

Exposure treatment consists of nine bi-weekly, 90-minute sessions. The first two sessions are devoted to information gathering, explanations of treatment rationale, and treatment planning. Clients are told that:

'It is extremely difficult to digest painful experiences and it takes a great deal of effort to deal with such experiences. Many of your assumptions or expectations about men, sex, or the world in general may have been shattered and you haven't had the opportunity to rebuild them. We're here to help you do that. Often the experience comes back to haunt you through nightmares, flashbacks, phobias, depression, etc., because it is "unfinished business".'

'What we are going to do is the opposite of our tendency to avoid discomfort. We'll help you to digest the experience by helping you to expose yourself to the assault in your imagination and by helping you to stay with it long enough to get more used to it. The fleeting images or thoughts about the rape that you do have, like flashbacks or nightmares, stop short of finishing the process when the intense fear or emotions make it too uncomfortable. We will help you use imagery to approximate the memory as closely as possible—not only seeing the attack in your mind, but reliving it with all the emotions and feelings you felt at the time. The goal is to be able to have these thoughts, to talk about the rape, or even see the cues associated with the rape without experiencing the intense anxiety that is disrupting your life.'

A hierarchy of avoided situations is constructed for *in vivo* exposure homework. Clients are instructed to try to imagine as vividly as possible the assault scene and describe it aloud. For the first two exposure sessions, clients are instructed not to verbalize details which are extremely upsetting. During the remaining sessions, they are encouraged to describe the rape in its entirety, repeating it several times for 60 minutes per session in order to facilitate habituation. The client's narratives are tape-recorded and they are instructed to listen to the tapes at home at least once daily. At the last session, clients are reminded that avoiding safe situations or rape-related thoughts may increase PTSD symptoms and are asked to practise this approach in their everyday life.

Supportive counselling follows the same nine-session format. Treatment focuses on assisting clients in solving daily problems which may or may not be rape-related and thus aims at promoting the perception of self-control. The therapist plays an indirect and unconditionally supportive role.

Preliminary analyses indicate that SIT and exposure result in PTSD symptom reduction more than supportive counselling or a no-treatment wait-list control.

Difficulties in treating rape victims

Many rape victims who suffer from PTSD and other rape-related problems show reluctance to seek treatment. Many have suffered in silence for years before presenting for treatment. There are several reasons for this. Many rape victims wish to forget this experience and therefore do not discuss it with others. Moreover, they tend to deny the possibility that the rape may have significantly impacted on their lives. They seem to adopt an avoidant coping style which may hinder adequate processing of the trauma and thus result in chronic difficulties.

When rape victims finally enter treatment, their avoidant coping style may be reflected in erratic attendance and a high frequency of non-compliance with homework assignments. The therapist must remember that for this client population, 'being a bad patient' may reflect the presence of post-traumatic stress disorder rather than resistance to treatment, lack of consideration, or manipulation. Our clinical experience indicates that a large proportion of those who seek treatment profit from it despite the difficulties mentioned above.

A special problem is posed by treating rape victims via repeated re experiencing of the trauma. Unlike simple phobics who can easily imagine their phobic situation and experience fear during imagery, many rape victims have difficulty emotionally reliving their traumatic experience during treatment. Since emotional response during imagery has been repeatedly found to predict treatment success, it is important to create a therapeutic atmosphere to help the client overcome her tendency to dissociate her emotions from the traumatic memory.

Clinical guidelines

1. Therapists should create an atmosphere that allows the victim to relate the details of her assault and to express her distress freely. Many victims, even years after the assault, burst into tears when recounting their assault.
2. Therapists should ensure that the patient is feeling more in control of her emotions before leaving the session.
3. If a victim manifests dissociative symptoms such as numbing or cognitive avoidance, treatment by exposure (as described above) is recommended.
4. If the victim manifests primarily chronic arousal (i.e. startle) and anxiety symptoms, SIT (stress inocualtion treatment) (as described above) is recommended.
5. If a victim shows the entire clinical picture of PTSD (post-traumatic stress disorder), a combination of exposure and SIT, beginning with SIT, is recommended.

Acknowledgements

Preparation of this manuscript was supported by NIMH Grant No. 5 RO1 MH42178–02 awarded to the first author.

References

American Psychiatric Association (1987). *Diagnostic and statistical manual of mental disorders—revised*. APA Washington, DC.

Ellis, A. (1977). A basic clinical theory and rational-emotive therapy. In *Handbook of rational-emotive therapy* (ed. R. Grieger), pp. 3–34. Springer, New York.

Foa, E. B. and Kozak, M. J. (1986). Emotional processing of fear: exposure to corrective information. *Psychological Bulletin*, **99**, 20–35.

Frank, E. and Stewart, B. D. (1983). Physical aggression: treating the victims. In *Behaviour modification with women* (ed. E. A. Bleckman), pp. 245–72. Guilford Press, New York.

Frank, E. and Stewart, B. D. (1984). Depressive symptoms in rape victims. *Journal of Affective Disorders*, **1**, 269–77.

Frank, E., Anderson, B., Stewart, B. D., Dancu, C., Hughes, C. and West, D. (1988). Efficacy of cognitive behavior therapy and systematic desensitization in the treatment of rape trauma. *Behavior Therapy*, **19**, 403–20.

Keane, T. M., Fairbank, J. A., Caddell, J. M., and Zimering, R. T. (1989). Implosive (flooding) therapy reduces symptoms of PTSD in Vietnam combat veterans. *Behavior Therapy*, **20**, 245–60.

Kilpatrick, D. G., Veronen, L. J., and Resick, P. A. (1982). Psychological sequelae to rape: assessment and treatment strategies. In *Behavioral medicine: assessment and treatment strategies* (ed. D. M. Dolays and R. L. Meredith), pp. 473–97. Plenum Press, New York.

Lange, A. J. and Jakubowski, P. (1976). *Responsible assertive behavior*. Research Press, Champaign, Il.

Resick, P. A., Jordan, C. G., Girelli, S. A., Hutter, C. K. and Marhoefer-Dvorak, S. (1988). A comparative outcome study of behavioral group therapy for sexual assault victims. *Behavior Therapy*, **19**, 385–401.

Rychtarik, R. G., Silverman, W. K., Van Landingham, W. P. and Prue, D. M. (1984). Treatment of an incest victim with implosive therapy: a case study. *Behavior Therapy*, **15**, 410–20.

Turner, S. M. (1979). Systematic desensitization of fears and anxiety in rape victims. Paper presented at the Association for the Advancement of Behavior Therapy, San Francisco, CA.

Veronen, L. J. and Kilpatrick, D. G. (1982). Stress inoculation training for victims of rape: efficacy and differential findings. Presented in a symposium entitled '*Sexual violence and harassment*' at the 16th Annual Convention of the Association for Advancement of Behavior Therapy, November 1982, Los Angeles, CA.

17

HIV and AIDS-related psychiatric disorder: what can the psychiatrist do?

JOSE CATALAN

Introduction

As the dust raised by the newspaper headlines and official information campaigns begin to settle, the true extent of the problem of HIV (Human Immunodeficiency Virus) infection and its likely spread in the UK is becoming apparent: by the end of 1987, between 20 000 and 50 000 individuals were thought to be HIV infected in England and Wales; by the end of 1992, between 10 000 and 30 000 cases of AIDS (Acquired Immune Deficiency Syndrome) are expected; over the next 10 to 15 years, the number of cases of AIDS is likely to be between 16 000 and 40 000 in England and Wales (HMSO 1988).

HIV infection, apart from its wide-ranging social and individual implications, is affecting the practice of medicine, both in developed and developing countries. HIV-related problems are present in most medical specialities, and renewed attention is being paid to almost forgotten infection control issues. Furthermore, old ethical and legal questions are being asked again in a new context.

Psychiatry is not being spared: clinical syndromes directly related to HIV infection of the central nervous system or secondary to the disease, as well as the understandable psychosocial reactions to a potentially fatal condition have been described. Issues concerning the management of patients with HIV disease (or suspected of it) continue to receive attention. Ethical and legal questions concerning confidentiality, consent to investigations, and treatment and detention in hospital, are being debated.

There are several reasons for the controversies concerning the psychiatric aspects of HIV infection. First, while information is accumulating rapidly, there is still a good deal of genuine uncertainty about the prevalence of psychosocial and neuropsychiatric syndromes in HIV disease and about their treatment, course, and significance. Second, a rational approach to HIV disease is often prevented by emotive and irrational reactions caused by fears of contamination, disapproval of the lifestyle of patients, or identification with the 'victim'—against the background of a condition which stigmatizes both those suffering from it and their careers. Finally, few psychiatrists in the UK (outside metropolitan areas) have so far had much contact with HIV patients; such lack of familiarity, even assuming an unprejudiced approach, can mean that some of the practical and ethical issues involved have not been thought through adequately. The way ahead towards a resolution of the controversies will, at the very least, demand of psychiatrists: (1) the acquisition of accurate theoretical and practical knowledge, and (2) the development of positive and realistic attitudes to HIV patients and their relatives.

In this chapter an outline will be given of the psychological and neuro-psychiatric problems of HIV infection, followed by a review of some areas of practical difficulty and a discussion of guidelines to deal with these difficulties.

Psychosocial and neuropsychiatric problems in HIV infection

Only an overview of psychiatric syndromes in HIV infection will be pro-vided here. More detailed information about them (Dilley *et al.* 1985; Catalan 1988; Goldmeier and Granville-Grossman 1988; Rosenblum *et al.* 1988; WHO 1988) and about HIV infection in general (Gottlieb *et al.* 1987) is availale in recent publications. In general, it should be remembered that most of the literature on the psychiatry of HIV disease comes from developed countries and it is mainly about gay/bisexual men; it would seem premature to assume that these findings can be universally applied. More reliable data should become available as the results of comprehensive prospective studies, involving patients from different transmission groups (haemophiliacs, in-travenous drug users (IVDUs), and others) are published, but these are still some way away. In the meantime, it needs to be recognized that detailed information about the prevalence of psychiatric syndromes is limited.

Psychological reactions (adjustment reactions, neurotic disorders, de-pressive disorder) and organic brain syndromes (acute and chronic) can occur in HIV patients. The type of disorder and its prevalence will be influenced by such factors as the stage of the disease, personality character-istics and individual vulnerability, and the extent of social and psychologi-cal supports.

Psychosocial problems

Many individuals experience adjustment reactions on learning of a positive HIV antibody test result, with symptoms of anxiety/depression and in-somnia, but these disorders are usually self-limited, and more persistent syndromes are less frequent. Fluctuations in symptoms are common, usually in response to health and social changes. The way in which the test result is disclosed, and the degree of preparation before and support after the test will affect the impact of the disclosure. While depressive symptoms are common, severe depression is less so, but there are reports of suicide following notification of a positive test result, and it is therefore important to be aware of this risk and ensure that effective help is provided.

Some people show high levels of anxiety and fear of HIV infection, in spite of one or more negative test results. Morbid fears of infection may be symptoms of hypochondriasis, or secondary to another condition, such as severe depression or a paranoid state.

Psychiatric problems are more common in individuals with physical symptoms related to HIV disease progression, and studies have shown that patients with persistent generalized lymphadenopathy (PGL) and AIDS-related complex (ARC) have greater psychiatric morbidity than patients who have actually developed AIDS (i.e. HIV infection in the presence of pathognomonic opportunistic infections, cancers such as Kaposi's sarcoma and lymphoma, or other complications), who often have long periods without physical problems.

The psychological problems which follow the diagnosis of AIDS are similar to those seen in other fatal conditions, but here they are magnified by the stigma attached to the disorder and by fears of rejection by family and others, which often lead to social isolation and withdrawal from potential supports. The prevalence of psychiatric disturbance is increased shortly after the diagnosis is made and towards the later stages of the disease. Major depression, with increased suicide risk (Marzuk *et al.* 1988), and adjustment reactions, are common at these stages, while good adjustment is often found during periods of reasonable health.

Neuropsychiatric problems

HIV infection can be associated with a variety of neuropsychiatric and neurological syndromes resulting from the direct effect of HIV on the nervous system, or secondary to opportunistic infections (such as cerebral toxoplasmosis and cryptococcal meningitis), vascular disorders, and intracranial tumours (such as lymphomas and Kaposi's sarcoma). Nervous system manifestations can occur early in the disease, at seroconversion, taking the form of a transient acute aseptic meningitis, while chronic neuropsychiatric syndromes are more likely to develop as the infection progresses (Rosenblum *et al.* 1988).

Research evidence to date suggests that asymptomatic HIV antibody-positive individuals do not differ significantly from negative controls in their performance on traditional neuropsychological tests, but this issue is not yet absolutely settled, and further research will be needed before reaching definitive conclusions (see Tross *et al.* (1988) for a discussion of some of the methodological problems involved). There is evidence, however, that as HIV infection progresses and the patient becomes symptomatic, the likelihood of neuropsychological abnormalities increases.

A neuropathological picture of encephalitis attributed to HIV has been found in a majority of AIDS cases, and the term 'AIDS dementia complex' has been introduced to describe a chronic syndrome characterized in the early stages by cognitive and behavioural changes (including apathy, psychomotor retardation, impaired memory and concentration) and motor abnormalities, which tends to progress to global intellectual impairment with major neurological signs (Rosenblum *et al.* 1988). It is not clear what the exact prevalence of this syndrome and its course is. Studies of un-selected samples suggest that between 8 and 16 per cent of AIDS patients develop the disorder, while autopsy studies of AIDS patients referred to neurologists suggest a figure in excess of 60 per cent (WHO 1988). Neuro-logical and neuropsychiatric manifestations may be amongst the presenting features of AIDS, developing sometimes in the absence of other signs of disease. On the positive side, there is evidence that antiretroviral therapy (zidovudine) can have beneficial effects on neuropsychological perform-ance in AIDS patients (Schmitt *et al.* 1988).

Another area of debate concerns the development of psychotic illness, typically mania or schizophrenia-like disorders, in patients with HIV infec-tion. Symptoms and signs suggestive of an organic brain syndrome are not always found, so that the aetiology of the disorder can, at times, remain unclear. While these disorders could be manifestations of an organic brain process (secondary to HIV infection or its treatment, or to drug misuse, for example), the possibility of a chance association cannot be ruled out (Halstead *et al.* 1988; Vogel-Scibilia *et al.* 1988).

HIV disease and psychiatry: some practical problems

The above outline of the range of HIV-related psychiatric disorders points to a variety of clinical problems, which include diagnostic and management issues, as well as ethical and legal ones. We are far from being in a position of having categorical and definitive answers to many of these problems, and it is likely that there will be changes of emphasis and direction in the future.

When should the psychiatrist think about the possibility of HIV-related psychiatric disorder?

This is not an easy question to answer, as no single indicator is pathogno-monic. In a general sense, it could be argued that psychiatrists should always think about this possibility, if only to be able to reject it with some confidence. In practice, however, the overall prevalence of HIV infection in a particular geographical area, and specific local characteristics, such as the existence of a serious intravenous drug misuse problem, for example, will influence the likelihood of HIV patients coming into contact with the local psychiatric services.

Several factors should be considered in relation to the possibility of HIV-related psychiatric disorder.

The type of psychiatric syndrome   The variety of psychiatric presentations and their lack of specificity would make it very difficult to be certain on clinical grounds alone. Affective disorder, adjustment reactions, and organic brain syndromes, for example, are so common in everyday clinical practice, that their presence would be of little help. In addition, the still relatively low prevalence of HIV infection in the population would also lead to expect low rates of infection when relying on diagnosis alone. In a study involving 69 patients referred for investigation of dementia over a three-year period, none was found to have HIV infection (Sulkava *et al.* 1987), and Halstead and co-workers (1988) were able to document only five cases of HIV infection presenting with psychosis in an area dealing with over 2000 HIV patients.

Patients whose psychiatric presentation includes fears or delusions about HIV disease (as in cases of hypochondriasis, severe depression, or schizophrenia) are in fact unlikely to have the condition or to have been at risk (Miller *et al.* 1988).

The characteristics of the patient   Membership of one of the transmission groups for HIV could be regarded as an indicator of risk of infection. Thus, in the UK, gay and bisexual men, IVDUs, haemophiliacs, recipients of infected blood, and the sexual partners of these groups, would be considered to be at risk. Elsewhere these groups (African countries with high prevalence of heterosexual infection) or their relative proportions (Italy and Spain, where IVDUs are predominant amongst AIDS cases) may be different.

However, when dealing with individual cases, there are risks in putting too much emphasis on group characteristics: for example, singling out people belonging to any of these groups would lead to a high rate of 'false positives'—after all, only a minority of gay men, recipients of blood or blood products, IVDUs, and their sexual partners, are known to be infected. At the same time, the possibility of a significant number of 'false negatives' (those who had no obvious risk factors, or who deliberately concealed them) would add to the inaccuracy of this approach, if it were to be taken in isolation. Clearly, patient characteristics are important when considering whether a patient is likely to have HIV infection, but other factors, such as a history of specific behaviours likely to lead to infection, would need to be included.

History of behaviour likely to lead to HIV infection   Evidence of a history of behaviour known to be associated with the transmission of HIV (unprotected anal intercourse, intravenous drug use, sexual contact with individuals with HIV infection, receiving infected blood or blood products, etc.), would increase the likelihood of infection. However, some patients may be reluctant to reveal some of these activities, partly because of their own fears of infection, but also as a result of concern about rejection by doctors or other health workers. It follows that careful history-taking, with sensitive handling of questions about the risky behaviours would be essential both to obtain accurate information and to avoid alienating the patient. It will be particularly important for doctors to make sure that their own views about patients' lifestyles and behaviour do not interfere with their clinical judgement.

The physical condition of the patient   On its own, the presence of symptoms and signs compatible with immuno-deficiency (severe weight loss, malaise, chest infection, diarrhoea, neurological symptoms, etc.), would be too non-specific and therefore of little value, but together with other factors such a clinical picture may lead to the possibility of HIV disease.

The patient is known to have HIV disease   It should not be assumed that a psychiatric disorder in someone with HIV infection is necessarily caused or related to it. Clearly, a depressive adjustment reaction developing shortly after a patient has been given a positive HIV test result is likely to be causally linked to the impact of the disclosure. However, in other cases it may be harder to be certain of the particular factors involved, as in the case of a patient with HIV infection and a history of recurrent depression who presents with a further relapse, or in patients with a psychotic illness (Vogel-Scibilia *et al* 1988). The problems involved in dealing with patients with known HIV disease and psychiatric disorder are considered in more detail below.

What should the psychiatrist do when it seems probable that a patient suffers from HIV infection?

In this instance, the psychiatrist will be faced with a series of questions needing careful consideration. As before, there are no easy answers, and only guidelines can be given.

How would knowledge of the patient's HIV status affect assessment and treatment?   Most psychiatric disorders in HIV patients, especially in asymptomatic ones, are likely to be coincidental or only indirectly associated with HIV infection. In the majority, the treatment of the psychiatric disorder is therefore unlikely to be significantly affected by knowledge of the patient's HIV status: severe depressive disorders or paranoid states, whatever their aetiology, will receive similar symptomatic treament. For example, in the case of the widow of a haemophiliac who has died from AIDS, and who presents with a depressive disorder, establishing the woman's HIV status may not be the first priority.

Knowledge of a patient's HIV status will be important, however, in the case of organic brain disorders whether caused by the direct effects of HIV on the CNS (such as acute aseptic meningitis or dementia), or by problems related to immuno-deficiency (such as cerebral toxoplasmosis, intracranial tumours, or pneumonia), where treament of the cause, as well as the symptoms of the disorder, will be necessary.

The possible effects on others involved in the management of the patient need to be considered as well: is there a risk of infection to the patient's sexual partner, to other patients, or to staff, and would such risks be reduced if the HIV status of the patient were known? The answers to these questions are not as obvious as they might seem: for example, knowledge of HIV status is not essential for giving up risky activities (Miller *et al.* 1986); infection control in hospital settings is more likely to be effective if the guidelines used to minimize infection are applied to all patients, and not just to those thought to be at risk (Catalan *et al.* 1989).

In summary, establishing a psychiatric patient's HIV status would be desirable if knowing it would result in clear benefits to the patient in terms of his or her psychiatric and physical health, and lead to an identifiable reduction in the risk of infection to others.

If establishing a patient's HIV status is thought to be desirable, how should this be done? Testing for antibodies to HIV, whether at the suggestion of the doctor or at the request of the patient, should only be carried out with the patient's informed consent (other than in exceptional cases, see below), and after the patient has had the opportunity to consider carefully the implications of a positive test result. This advice is in line with the recommendations made by the General Medical Council (1988) and the Royal College of Psychiatrists (Catalan *et al.* 1989). As the General Medical Council points out, this is so 'not because the condition is different in kind from other infections but because of the possible serious social and financial consequences which may ensue for the patient from the mere fact of having been tested for the condition'. In practice, a patient is very unlikely to refuse the doctor's suggestion of testing for HIV when this is done as part of the assessment and treatment of the patient's condition.

The process of discussing the practical aspects and implications of HIV testing (usually known as 'pre-test counselling') can be very technical and requires a good deal of knowledge about HIV infection. This kind of counselling should therefore be carried out by trained and experienced staff, and so psychiatrists undertaking this task should ensure they have the necessary knowledge. It is also advisable to provide 'post-test counselling', whatever the test results.

Testing an individual for HIV, either at the person's request or at the suggestion of a doctor, should not be confused with *screening*, which is the systematic applications of HIV testing to establish the prevalence of infection in a particular population or target group, bodily product, etc. Screening could be carried out anonymously, with or without the patients' explicit consent, or in other forms. There is no place for routine screening in everyday psychiatric practice, and screening should only be carried out as part of a specific research project, with clear aims and methods, and after ethical and legal questions have been tackled (Catalan *et al.* 1989).

How should the patient be managed if he or she refuses to be HIV tested?
When recommending testing for HIV, the doctor should stress the likely benefits that would follow (adequate diagnosis and treatment of the psychiatric disorder, provision of early treatment for HIV infection, prevention of medical complications), as well as possible problems. The patient should be reassured about confidentiality and further support should be offered for the patient and, if appropriate, relatives. The patient's concerns and anxieties should be explored and tackled. If in spite of this the patient still refuses to be tested, the doctor will need to decide whether this is one of the exceptional cases where testing without the patient's consent should be carried out (see below). If the doctor concludes that the patient is capable of giving consent, and that testing is not (in the words of the General Medical Council) 'imperative in order to secure the safety of persons other than the patient, and where it is not possible for the prior consent of the patient to be obtained', the patient's wishes should be respected.

In terms of general management, in particular as regards nursing and other procedures, the patient should be managed as if he or she were HIV positive. This should not present any significant practical problems, especially if adequate infection control procedures are applied to all patients (see below).

HIV testing without consent in psychiatric practice   Testing without the patient's explicit consent would only be justified in very rare instances. In line with the recommendations of the General Medical Council (1988), this would only occur when testing was essential to avoid infection to others. In psychiatry, however, this issue is likely to arise also in cases where, as a result of psychiatric illness, the patient is either unwilling to give consent (for example, a manic patient), or unable to do so (because of severe depression or cognitive impairment). Psychiatrists are familiar with these issues in the context of the Mental Health Act 1983, but the provisions of the Act do not cover all the possible problems arising in relation to HIV infection.

Faced with the question of testing without the patient's consent, the psychiatrist should ensure that the reasons for testing are clearly established: in particular, in what way the patient's treatment would be influenced by knowledge of HIV status and what benefit to the patient is to be expected; to what extent it would ensure the safety of staff or other patients, and could such safety be achieved in the absence of knowledge of the patient's test result?

A decision to test without the consent should only be reached after discussion with the rest of the clinical team and, ideally, in consultation with a physician with experience of HIV disease. The reasons should be clearly stated in the notes. It has been suggested that the doctor's legal advisor should be involved. In general, the doctor making this decision should be prepared to defend it in court if necessary (Catalan *et al.* 1989).

How should the known HIV patient be managed?

Assessment and management of the psychiatrist disorder   When a known HIV patient is referred for psychiatric assessment, careful consideration will need to be given to the possible contribution of the infection to the psychiatric syndrome (see above). This is in no way different from the assessment problems that arise when dealing with other patients with both psychiatric and physical disorders. A common reason for referral of HIV patients to psychiatrists is for assessment of low mood and lack of energy and concentration. In such cases, the relative contribution of a depressive disorder, early cognitive impairment related to an organic brain syndrome, and the tiredness and weakness associated with loss of weight and opportunistic infections, amongst others, will need to be considered.

A full physical examination, including neurological assessment, will be required. Discussion with the patient's physician to obtain information about the patient's degree of immuno-deficiency and past medical history will be appropriate. Neuropsychological assessment and neuroradiological investigations would be indicated when an organic brain syndrome is suspected. Such intensive assessment is important to establish the cause of the disorder and to provide adequate treatment: opportunistic infections and other problems can be treated effectively, as can functional psychiatric disorders. Symptomatic physical treatments can be used as in other psychiatric disorders, but it is recommended that low dosages and drugs with few side-effects are used, to avoid the risk of complications and adverse reactions (Ostrow *et al.* 1988).

Who should be informed of the patient's HIV status?   Like any other patient, those with HIV should expect reasonable standards of confidentiality. However, there are some limits to medical confidentiality, and patients should be aware of them. Patients are usually treated by a team, and so those members of the team who need to know the patient's status should be informed. Such a need to know would arise if it were essential to offer support to a depressed patient or the relatives, when the patient's condition is clearly related to concern about HIV status. The need to know would also arise if there were risks of infection to staff, as in the case of nurses dealing with a patient with severe diarrhoea and vomiting. In all these situations, it is important for the doctor to discuss the reasons for disclosure with the patient.

In many cases, it would be desirable to discuss the patient's condition with others, such as the patient's general practitioner or other doctors, and it is important that the patient's consent is obtained before doing so (General Medical Council 1988).

As for informing the patient's spouse or other sexual partner, the same principles should apply, so that disclosure without the patient's consent would only be justifiable when there is a 'serious and identifiable risk to a specific individual who, if not so informed, would be exposed to infection' (General Medical Council 1988).

Should any precaution be taken to ensure that other patients and staff do not become infected? The risk to other patients and staff in psychiatric and medical settings is minimal and it is limited to a few specific situations (parenteral inoculation with infected blood or body fluids, or contamination of lacerated skin or mucous membranes). It is essential that staff working in psychiatric units receive adequate training in infection control methods in relation to HIV disease.

In terms of infection control, the most effective way of reducing the risk of infection with HIV or other pathogens is to regard *all* patients as potentially capable of transmitting and acquiring infections. This approach will not only minimize the risk, but will also help to maintain confidentiality about infected patients.

Nursing an HIV patient in a single room is not essential from the infection control point of view, unless the patient is immuno-compromised and at risk of other infections from others, or if there is severe diarrhoea or bleeding. However, in some cases the patient's right to privacy and psychiatric needs may be better served by nursing in a single room. Ordinary social contacts, sharing a room or crockery, cleaning, serving food, and bed making, do not carry a risk of infection.

HIV patients who are aggressive or violent, or who show sexual disinhibition, present particular problems. Staff anxieties are usually raised by such patients, and team discussion of the problems, and a policy to deal with such patients will be necessary. The use of the Mental Health Act will need to be considered, as well as the use of psychotropic medication and nursing in secure facilities.

The role of the psychiatrist in supporting health care staff
Medical and nursing staff dealing with HIV patients often show evidence of psychological stress and other difficulties (Ross and Seager 1988). This is not surprising, as studies or staff stress in health workers involved in the

care of patients with advanced cancer or other terminal diseases have revealed similar findings (Vachon *et al* 1978; Wooley *et al* 1989)

Psychiatrists and other mental health workers are likely to play an important role in helping doctors and nurses deal with the stress associated with long-term contact with individuals suffering from HIV infection and their families. The risk of 'burnout' (Mayou 1987) may be reduced by means amongst other things of regular meetings of health workers involved in frequent face-to-face contact with HIV paients, preferably in a group setting.

Clinical guidelines

1. The psychiatric implications of HIV disease are likely to face psychiatrists with a range of dilemmas in relation to diagnosis, treatment, organization of services, and ethical and legal questions. While many of the issues are not new, the technical complexities and practical problems raised by HIV disease will make it necessary for psychiatrists to acquire adequate information to ensure optimum care is given to patients, and to work through possible emotional obstacles to dealing with HIV patients.

2. The need to be well informed about the psychiatric effects of HIV infection, and its legal and ethical aspects, will be apparent when considering whether a psychiatric patient is likely to be suffering from HIV infection, and the steps to follow to establish the diagnosis. In particular, psychiatrists need to be aware of issues concerned with investigation of HIV infection and testing for HIV, and the rare situations where testing without the patient's consent might be considered acceptable.

3. HIV disease highlights the need for collaboration with other professionals such as physicians and psychologists, in the process of diagnosing and treating HIV-related psychiatric disorders.

4. Psychopharmacological treatments have a place in HIV disease, but need to be used with care in view of the likely risk of adverse reactions and side-effects.

5. Psychiatrists have a potentially important role in supporting individuals involved in face-to-face care of HIV patients, and in the development and evaluation of methods of staff support.

Acknowledgements

I am indebted to Jasmine Webster for her prompt and impeccable secretarial assistance.

References

Catalan J. (1988). Psychosocial and neuropsychiatric aspects of HIV infection: review of their extent and implications for psychiatry. *Journal of Psychosomatic Research*, **32**, 237–48.

Catalan, J., Riccio, M. and Thompson, C. (1989). HIV disease and psychiatric practice. *Psychiatric Bulletin*, **13**, 316–48.

Dilley, J., Ochitill, H., Perl, M. and Volberding, P. (1985). Findings in psychiatric consultations with patients with AIDS. *American Journal of Psychiatry*, **142**, 82–5.

General Medical Council (1988). *HIV infection and AIDS: the ethical considerations*, General Medical Council, London.

Goldmeier, D. and Granville-Grossman, K. (1988). The clinical psychiatry of HIV infection. In *Recent advances in clinical psychiatry, No. 6* (ed. K. Granville-Grossman) pp. 1–49. Churchill Livingstone, Edinburgh.

Gottlieb, M. S., Jeffries, D. J., Mildvan, D., Pinching, A. J., Quinn, T. C., and Weiss, R. A. (eds) (1987). *Current topics in AIDS, Volume 1*. Wiley, Chichester.

Halstead, S., Riccio, M., Harlow, P., Oretti, R. and Thompson, C. (1988). Psychosis associated with HIV infection, *British Journal of Psychiatry*, **153**, 618–23.

HMSO (1988). *Short-term prediction of HIV infection and AIDS in England and Wales*. HMSO, London.

Marzuk, P. M., Tierney, H., Tardiff, K., Gross, E. M., Morgan, E. B., Hsu, M. and Mann, J. J. (1988). Increased suicide risk in persons with AIDS. *Journal of the American Medical Association*, **259**, 1333–7.

Mayou, R. (1977). Burnout. *British Medical Journal*, **295**, 284–5.

Miller, D., Acton, T. and Hedge, B. (1988). The worried well: their identification and management. *Journal of the Royal College of Physicians*, **22**, 158–65.

Miller, D., Jeffries, D., Green, J., Willie Harris, J.R., and Pinching, A. (1986). HTLV-III: should testing ever be routine? *British Medical Journal*, **292**, 941–3.

Ostrow, D., Grant, I. and Atkinson, H. (1988). Assessment and management of the AIDS patient with neuropsychiatric disturbances. *Journal of Clinical Psychiatry*, **49**, 14–22.

Rosenblum, M. L., Levy, R. M., and Bredesen, D. E. (eds) (1988). *AIDS and the nervous system*. Raven Press, New York.

Ross, M. and Seager, V. (1988). Determinants of reported burnout in health professionals associated with the care of AIDS patients. *AIDS*, **2**, 395–7.

Schmitt, F. A., Bigley, J. W., McKinnis, R., Logue, P. E., Evans, R. W. and Drucker, J. L. (1988). Neuropsychological outcome of AZT treatment of patients with AIDS and ARC. *New England Journal of Medicine*, **319**, 1573–8.

Sulkava, R., Korpela, J., and Erkinjuntti, T. (1987). No antibodies to HTLV-I and HIV in patients with dementia in Finland. *Acta Neurologica Scandinavica*, **76**, 155–6.

Tross, S., Price, R. W., Navia, B., Thaler, H. T., Gold, J., Hirsch, D. A. and Sidtis, J. J. (1988). Neuropsychological characterization of the AIDS dementia complex: a preliminary report. *AIDS*, **2**, 81–8.

Vachon, M., Lyall, W. and Freeman, S. (1978). Measurement and management of stress in health professionals working with advanced cancer. *Death Education*, **1**, 365–75.

Vogel-Scibilia, S. E., Mulsant, B. H. and Keshavan, M. S. (1988). HIV infection presenting as psychosis: a critique. *Acta Psychiatrica Scandinavica*, 78, 652–6.

World Health Organization (1988). *Report of the consultation on the neuro-psychiatric aspects of HIV infection*. Global Programme on AIDS, WHO, Geneva.

Wooley, N., Stein, A., Forrest G. and Baum, J. (1989). Staff stress and job satisfaction at a children's hospice. *Archives of Diseases in Childhood*, 64, 114–18.

10

Psychosurgery: is it ever justified?

JOHN COBB and DESMOND KELLY

Introduction

The first account of psychosurgical intervention in man was published nearly a hundred years ago (Buckhardt 1891). The outcome in that case was poor, and little more was done until the neurophysiological studies of Fulton and Jacobson (1935) inspired the Portuguese neurosurgeon Egaz Moniz to operate on 20 patients suffering from irrational fears, severe anxiety, and crippling obsessional behaviour. His enthusiastic report was influential and followed by a rapid increase in psychosurgical treatment. The suffering caused by schizophrenia before the advent of phenothiazines was enormous, and once a person had been in hospital for a year, it was likely that they would remain there for the rest of their life. By 1950 the Americans Freeman and Watts had operated on over 1000 patients and the estimated world-wide figure for such treatments was 20 000 (Clare 1980).

Therapeutic zeal far exceeded objective, scientific evaluation. Various 'free-hand' neurosurgical techniques were introduced, which were then applied in an uncontrolled manner on psychiatric patients with a wide range of diagnoses. The characteristics shared by these disorders were severity and chronicity. Inevitably this unbridled enthusiasm produced a strong negative reaction. Destruction of central nervous system tissue in which no anatomical or physiological abnormality can be demonstrated is difficult to accept, leading to antagonism based on aesthetic rather than scientific objections. A furious debate ensued, inevitably generating much more heat than light. This occurred when powerful new behavioural methods were being developed for the treatment of obsessive-compulsive neurosis (Marks 1981, pp. 45–95). The number of operations performed in the United Kingdom fell dramatically from an average of 1100 operations a year during 1948–54, to 119 in 1976 (Clare 1980), and to 37 in England and Wales in 1983 (Hussain et al. 1988). Ironically, while this furore was at its height, neurosurgeons were poineering new precise stereotactic techniques, and clinicians were publishing studies which systematically evaluated outcome of precisely defined operations in distinct groups of patients (Tan et al. 1971; Mitchell-Heggs et al. 1976).

Psychosurgery in 1989 arouses much less controversy than ten years ago, but this must be at least in part due to the fact that it is only rarely carried out. However, a recent survey in two regions of Britain showed that a majority of senior psychiatrists wished to retain and perhaps improve upon facilities for psychosurgery (Snaith *et al.* 1984). An important question must be, 'Have we, because of political and social pressure, abandoned an effective treatment which can bring relief to the suffering of chronically disabled patients?'.

Obsessive-compulsive neurosis

Background factors

Psychosurgery now is largely limited to the treatment of intractable, life-threatening depression and obsessive-compulsive neurosis. This chapter focuses mainly on obsessive-compulsive neurosis, a circumscribed condition, largely free of diagnostic ambiguity and controversy. The authors have had a special interest in the treatment of this condition dating back 20 years. Over 60 patients with severe, chronic obsessional illness have been treated over the past 20 years with a standard procedure (stereotactic limbic leucotomy). Systematic evaluation using standard measurements as well as regular clinical assessment allows most of the contentious issues which have been raised concerning psychosurgery to be examined. Controlled, double-blind treatment studies are both impractical and unethical. However, in a condition such as obsessional neurosis which is known to be chronic, with a spontaneous remission rate of 20 per cent (Kringlen 1970), and a low rate of response to 'placebo' treatments (Rachman and Hodgson 1980, pp. 311–14), this is less of an objection than with the treatment of other psychiatic conditions.

Experimental basis

Stereotactic limbic leucotomy (Kelly 1980) is based both on animal and human studies which have led to certain assumptions, some more soundly based than others.

1. The limbic system plays a central role in the generation of emotion, which particularly in humans is modified and modulated by the frontal lobes (Kelly 1980). Within the limbic system the Papez circuit, including the hippocampus, fornix, mammillary bodies, anterior nucleus of the thalamus, and cingulate gyrus, plays an important role in the integration of memory and emotion. It may also serve to modulate the effects on higher mental processes of phylogenetically ancient influences emanating from 'reptilian' visceral brain. The anterior part of the Papez circuit is primarily involved in emotion and the posterior part in memory.

2. Both in the anterior cingulate gyrus and in the lower medial quadrant of the frontal lobe, fibre tracts involved in emotions are very closely associ-

ated with autonomic pathways. Thus even under general anaesthesia emotional pathways can be identified by looking for areas which, when stimulated, produce physiological change (Kelly 1980).

3. Concentration of fronto-limbic connections are localized in three main areas:

    (a) the anterior part of the cingulate gyrus;

    (b) the lower medial quadrant of the frontal lobe; and

    (c) fibre tracts overlying the posterior orbital cortex.

Both (b) and (c) overlap and converge posteriorly.

4. Results from a number of different, less precise surgical procedures indicate that lesions made in the anterior cingulate region and lower medial quadrant of the frontal lobe are followed by relief of symptoms in patients suffering from anxiety and obsessional neurosis (see Table 10, Yaryura-Tobias and Neziroglu 1983).

Table 10 *Clinical outcome after psychosurgery for obsessive-compulsive neurosis. The Table is based on 11 studies (Yaryusa-Tobias and Neziroglu 1983), and includes results from both stereotactic and free-hand operations*

Number of patients	324
Symptom free	71 (28%)
Much improved	110 (33%)
Moderately improved	79 (24%)
Slightly improved	28 (8%)
No change	46 (14%)

Procedure

Stereotactic limbic leucotomy This operation is performed under general anaesthesia (Richardson 1973; Kelly 1980). The method employs X-rays to outline the skull and the ventricular system in relation to a probe guidance system. After appropriate calculations, a stimulating electrode and then a lesion-making probe can be introduced to any point within the brain to the three-dimensional accuracy of 1 mm. After stimulation studies have been carried out, small lesions 8 mm in diameter are made in the lower medial quadrant of the frontal lobe, interrupting some fronto-limbic connections, and in the cingulum bundle, interrupting part of the Papez circuit. Stimulation produces physiological changes, in particular apnoea. On each side of the brain three 8 mm lesions are produced in the lower medial quadrant and two in the cingulate gyrus on each side.

Outcome

Stereotactic limbic leucotomy produced good outcome results in 49 patients with severe obsessive illness (Mitchell-Heggs *et al.* 1976). Twenty months following the operation, 49 per cent of patients were either much improved or symptom free and a further 29 per cent showed significant improvement. Improvement figures were even better at 20 month follow-up. In evaluating the significance of these results, it must be remembered that these patients had suffered severe symptoms for an average of 13 years, were incapacitated in their work ability and were disabled in their family and social relationships. Outcome was assessed openly using standardized assessor-rated and patient-rated instruments, in addition to clinical ratings based on interviews with key relatives as well as the patients themselves.

Pattern of improvement   Most patients show little change in the first two weeks. Many express disappointment that the hoped for 'miracle' has not occurred. Definite progress which is acknowledged by the patient may not be apparent until six weeks after the operation. Typically this is followed by a gradual improvement, accompanied by marked fluctuations over the next year. During this period behavioural and cognitive psychotherapy involving both patient and family is imperative. As well as focusing on specific obsessive rituals and ruminations, wider issues involving the patient's personal life and family dynamics usually need attention.

Illustrative case   LP was a 37-year-old woman who came from a family with no history of psychiatric problems. Her difficulties started in childhood. At the age of five years she developed a morbid fear of soiling herself and had to spend up to an hour in the toilet before going to school. By the time she was 13 this had become so severe that she was unable to attend school at all and had to have home tuition. At the age of 15 years she was admitted to the care of an eminent psychiatrist who treated her with antidepressant medication and ECT under modified narcosis. This improved her to the point where she was able to resume her education and take up a reasonably active social life. She qualified as a teacher and started work at a primary school. Her father died when she was 19 years old, following which she relapsed. This time she responded to clomipramine. She remained reasonably well for the next ten years but towards the end of this period her symptoms gradually returned. She had to give up her job and was unable to work until her operation nine years later.

The symptoms that she suffered included a wide range of obsessions and compulsions. Fears concerning contamination with faeces or urine which she had experienced as a child reappeared and forced her to spend about half an hour in the lavatory each time she wanted to pass urine. She had a number of associated cleaning and decontamination rituals, which involved both washing and bathing, and made these activities take at least three times as long as normal. There was also a fear that if she did not do things 'properly' then somebody would suffer serious harm. Some of these rituals involved checking household items such as taps and electrical appliances. Others involved bizarre 'magical' and superstitious ideas, which meant that virtually everything she did had to be done in certain sets of numbers. Everyday activities were disrupted. For example, she had to drink a cup of tea carefully counting the sips and making sure she got the numbers right. Both the number 13 and the colour green became taboo and this resulted in a wide range of practical difficulties.

A great deal of therapeutic energy went into trying to help LP. She saw four consultant psychiatrists and was treated with a wide range of physical treatments. This included adequate courses of clomipramine, monoamine-oxidase inhibitors, combined antidepressants, lithium, and ECT. Behaviour therapy was carried out by a qualified and experienced nurse therapist on a domiciliary basis over a three-month period. The therapist spent about 25 hours with her and in addition her mother was involved as a co-therapist. Other psychological treatment included psychotherapy from an Eriksonian psychotherapist and a course of treatment from a well-known therapist and author of books on self-help for anxiety. None of this treatment produced any lasting benefit.

After stereotactic limbic leucotomy her recovery followed a typical course (see Table 11). She showed little change immediately after the operation but then followed a fluctuating though overall steadily improving course over the next 15 months. She took her own discharge from hospital only 14 days post-operatively, declining the usual post-operative behaviour therapy because she said she already had plenty of knowledge about this treatment and would do the treatment herself! Three months after the operation she reported that she was steadily overcoming her obsessions one by one and had started to work in the afternoons as a nanny. Three months later she had lost most of her obsessions though she was still feeling anxious and finding it difficult to relax. She started to work in a sports shop doing both book-keeping and delivering goods. Nine months post-operatively she became rather depressed though her obsessions did not return. Her depression responded to 50 mgs clomipramine and subsequently to 25 mgs amitriptyline. By a year after the operation she had taken a job in a cafe working five days a week. She was starting to socialize again and was generally feeling much more relaxed and happy. By 18 months after the operation she was attending two further education classes and was actively looking for more exacting employment.

She talked with sensitivity about the way in which her appreciation of music and nature had increased. Her ability to enjoy herself and to form relationships was enhanced, but not at the cost of any loss of self-control or

Table 11   *Scores on mood and obsessional questionnaires for LP before and after stereotactic limbic leucotomy*

	Pre-op.	1/12 Post-op.	3/12 Post-op.	18/12 Post-op.	2 yrs Post-op.
Hamilton Depression Scale	31	3	9	8	9
Beck Depression Inventory	30	15	2	2	9
Maudsley Obsessional Checklist	30	27	20	4	9
Leyton Obsession Inventory:					
resistence	98	70	19	5	10
interference	68	66	5	1	4

disinhibition. Her mother commented, 'she is just better, balanced, and now she is no longer tortured by her obsessions. She is starting to lead a normal life'.

Implications of experience with obsessive-compulsive disorder

This chapter has so far focused on one specific psychosurgical operation in a well-defined diagnostic group. Put another way, we are studying the effects of anatomically precise lesions in a circumscribed psychopathological condition. This makes it easier to consider the key questions 'Do specific lesions produce beneficial results?' and if so, 'What is the underlying mechanism?' Much previous debate has been confused because a heterogeneous group of surgical procedures were applied either to a diagnostically heterogeneous group of patients or to diagnostic categories such as depression or schizophrenia which are less clearly defined clinical entities than obsessional neurosis.

Are controlled outcome studies possible?   The first general point that must be considered concerns the relation between clinical practice and the results of well-designed, controlled treatment studies. No one would dispute that medical treatment should be based on sound evidence which in general means well-controlled and replicated clinical trials. Setting up a controlled study of psychosurgery involves insuperable problems. The 1983 Mental Health Act insists that the patient has exact knowledge of what is to be done and gives full informed consent to the procedure. It is difficult to imagine asking a patient to assent to an anaesthetic and burr holes only. In any case the procedure that followed would have to be open rather than blind for legal reasons. Even prior to the 1983 Act, the Royal College of Psychiatrists had abandoned a planned controlled study of psychosurgery following ethical objections. Purists may insist that without a controlled clinical trial, treatment should be abandoned. Those faced with the responsibility of day to day care of seriously disabled patients, aptly termed '*The captive patient*' (Hunter Brown 1980), may take a more flexible view and be prepared to scrutinize single case studies in combination with systematic though controlled observations of groups of patients.

Relevance of case studies The case presented is a typical example of a patient with primary obsessive-compulsive neurosis for whom stereotactic limbic leucotomy appeared to produce dramatic improvement. She had been disabled by continuous symptoms over several years. Depression was part of her clinical picture, and earlier response to antidepressant treatments raises the question of whether the diagnosis is that of a primary obsessive-compulsive disorder or an underlying depressive illness. Obsessional symptoms dominated the clinical picture in the latter years of her illness and all four psychiatrists involved diagnosed primary obsessional neurosis. Adequate courses of a wide range of psychotropic drugs had been tried without benefit. The patient lived with her mother and thorough assessment of the home situation was followed by psychotherapeutic treatment involving the patient together with some conjoint sessions with her mother. Though conflicts were identified and discussed, obsessional symptoms were not altered. Behaviour therapy, involving exposure *in vivo* and response prevention, given by an experienced and specially trained therapist had produced only a fleeting improvement. The therapist commented that in her view treatment failed because challenging the rituals and ruminations generated very high levels of anxiety which did not habituate in the normal way. Thus, although the patient complied well with therapy and carried out the tasks set by her therapist with determination and courage, she was unable to tolerate the discomfort generated which to her appeared endless and unchangeable. This in the end made further intensive behaviour therapy unbearable for the patient. In-patient treatment was refused because following earlier admissions the patient's phobias extended to psychiatric hospitals. This is in marked contrast to the situation three months after the operation. Then she reported that on her own she was able to put into practice the exercises which her therapist had taught her nine months before, without too much anxiety, and that she was able to see definite signs of progress as a result.

Possible mechanisms of change with psychosurgery

If one accepts the evidence from this patient and from the rest of the series of 49 similar patients, together with that from other workers who have reported equivalent results with similar procedures (Mindus *et al.* 1985; Yaryura-Tobias and Neziroglu 1983), that improvement follows psychosurgery, one must next ask 'What is the mechanism of this change?'

Possible placebo response   Admission to hospital for psychosurgery is a dramatic business. Pre-operative preparation involves a range of investigations and procedures. These now include the stressful requirement to appear in front of a Mental Health Tribunal, which has the power to veto decisions to operate, however much the patients, their relatives and their psychiatrists want to proceed. This decision leaves no possibility of appeal. Thus the stage is set for a possible placebo response.

However, there are several arguments against a placebo mechanism. First, despite careful pre-operative counselling, many patients expect a dramatic cure, that they will wake after the operation free of their problem. Disappointment is inevitable when they find it is not so. Many patients go through a difficult period, which we describe as 'hitting the wall', about 2–10 weeks post-operatively when they discover that there has not been a 'magic cure'. In nearly all patients improvement occurs gradually with marked fluctuations, three steps forward and two steps back, over the course of a year or so. This pattern of change is quite unlike a placebo response.

Second, obsessive-compulsive patients are known to be poor placebo-responders (Rachman and Hodgson 1980, pp. 311–14). The patient, LP, gained no benefit from seeing an internationally known expert, even though having read the books of this therapist she strongly believed the treatment would work and went to great pains to organize the consultation for herself.

Third, patients report that after the operation they are less upset by their rituals and ruminations, though these may still occur as before. LP commented, 'I seem to be able to distance myself from them, and they don't go on and on endlessly in the same way'. Sometimes staff and relatives notice a definite change, with the patient looking more relaxed, though the patients themselves in the first two or three weeks will say that they still feel the same.

Postulated neurophysiological changes   If the operation works, as has been suggested, by interrupting reverberating circuits within the limbic system and its connections with the 'reptilian' brain on the one hand and with the frontal cortex on the other, these clinical observations are what might be expected. It may be postulated that obsessional compulsions and ruminations have their origin in the 'reptilian' brain. The drive from these areas is reinforced and intensified by associated emotional arousal in the limbic system. Furthermore these drives are given 'meaning' by the neocortex. Immediately after the operation the need to ritualize, emotional arousal, and associated compulsive thoughts are all still present and experienced by the patient. However, the association between these elements has been reduced and thus the vicious cycle which previously ensured their maintenance has been broken. Time is needed for the brain to reorganize its activities and for a new modulated balance between different levels of brain function to be established.

Adverse effects   Such an explanation also helps us to understand how it is that ten small discrete lesions can be made in physiologically active and important centres in the brain without producing any detectable deleterious effects. There is no doubt that the former free-hand frontal leucotomies could produce dire permanent effects on the personality. However, the patients in our series have been carefully scrutinized for a minimum of 18 months and much longer in many cases. It is exceptional for adverse changes in personality to be found. This is confirmed by sensitive observations based on clinical interviews, self-reports and accounts from relatives.

Patients do report persistent reduction in the duration of sleep. This is not associated with increased tiredness during the day, and appears to the patient as though the need for sleep has decreased. Often no complaint is made and the change is only revealed by direct questioning. A second change concerns sensitivity to medication. Patients are often more sensitive to side-effects of drugs post-operatively and require lower doses of preparations such as tricyclics to produce a therapeutic response. Sedatives (e.g., benzodiazepines) and alcohol have increased potency, which can lead to problems especially if the patient has had a previous tendency to substance abuse.

Psychosurgery in depression

Results of psychosurgery in obsessive-compulsive neurosis must be seen in the context of the use of operative procedures in psychiatric problems as a whole. In a survey involving 431 operations done over a three-year period (Barraclough and Mitchell-Heggs 1978), 63 per cent of patients were suffering from refractory depression as compared with 7 per cent with severe obsessional illness. Systematic outcome studies have produced results similar to those described with obsessive problems. Thus Bartlett *et al.* (1981) described recovery in two-thirds of patients following a stereotactic procedure. Furthermore it has been possible to compare the outcome of two different procedures, stereotactic subcaudate tractomy (Knight 1965) with stereotactic limbic leucotomy. Both procedures produce good results, though optimum benefit in severe depression appears to be produced by the first procedure, whereas the second has the edge in treatment of obsessional patients.

In depression as in obsessions, psychosurgery is indicated for chronic or persistently recurrent conditions. Transitory response to ECT is considered a good prognostic sign, though psychosurgery is likely to be less disruptive and traumatic than frequently repeated or long courses of ECT. Severely suicidal patients, refractory to other treatments, may be considered for psychosurgery after shorter periods of illness, or where there is a significant risk of death or lasting change in the capacity to live normal lives.

Future research

Many questions about psychosurgery remain, some of which may be explored with the new brain imaging techniques. For example, recent PET scan studies have shown abnormal glucose metabolism in the frontal lobes of patients with obsessional neurosis, while cognitive and psychophysiological studies have also indicated cerebrally mediated focal abnormalities (Eves and Tata 1989). Post-operative EEG changes can be used as a prognostic indicator (Evans *et al.* 1981). This type of work, combined with post-operative MRI scans, gives the possibility of linking the sites of operation with focal areas of dysfunction in the brain. Modern scanning techniques allow more precise location of lesion sites, and further comparative work may help elucidate the question as to which operation is best for which condition.

Medico-legal issues

On the political side, the impact of the 1983 Mental Health Act has not changed our selection procedure in any way. On the positive side, Mental Health Tribunals protect psychiatrists by providing an independent team consisting of another experienced general psychiatrist, a lawyer, and a lay person to re-evaluate the selection process. On the negative side, the bureaucratic process in setting up a tribunal is time-consuming and gives non-medical people and non-specialist psychiatrists power to contradict the clinical judgement of psychiatrists with greater experience, and to interrupt care, without their having to take any responsibility for the patients' future management. On balance we welcome the advent of the Mental Health Act Tribunal.

Ironically the Mental Health Act may be inadvertently discouraging the practice of psychosurgery, at a time when its value has been re-assessed and is more widely recognized. On the basis of the evidence and discussion in this chapter we would argue that in severe, intractable obsessive-compulsive disorder and in severe depression, psychosurgery should be less of a desperate last resort. Recognition of the potential benefits, leading to earlier consideration of surgery could prevent, as in the case presented, long periods of suffering and 'lost' years in young patients who should be enjoying the prime of life.

Clinical guidelines

Obsessive-compulsive neurosis

1. Uncontrolled data support the value of stereotactic limbic leucotomy (Kelly–Richardson operation) in obsessive-compulsive neurosis.

2. Clinical studies have failed to demonstrate serious, long-term side-effects of this operation.

3. The operation is only indicated in severe, chronic illness in which following criteria apply:

(a) the patient has been severely disabled by symptoms, for a period of at least five years (usually longer);

(b) treatment with adequate courses of pharmacological treatments, using conventional antidepressants alone and in combination, have failed;

(c) thorough courses of behavioural psychotherapy based on exposure and response prevention, carried out either with the patient as an-inpatient or on a domiciliary basis (preferably both) have failed;

(d) psychotherapeutic investigation of both personal and family problems, with intervention where appropriate, has been carried out;

(e) there is an absence of risk of drug abuse or alcoholism;

(f) there is also an absence of a history of significant violent or antisocial behaviour;

(g) co-existing organic cerebral pathology has undergone careful neuro-surgical investigation before proceeding to therapy;

(h) written approval of the Mental Health Commission (under the 1983 Mental Health Act) has been obtained.

Severe depressive illness

4. In severe, chronic, or persistently recurrent depressive illness, stereo-tactic subcaudate tractotomy appears to be the treatment of choice. This procedure should also be considered in more acute illness where the risk of suicide is great and there is no response to conventional treatment.

5. Age by itself is not an absolute contraindication to surgery, although the physical operative risks increase with age.

References

Barraclough, B. M. and Mitchell-Heggs, N. A. (1978). Use of neurosurgery for psychological disorder in the British Isles during 1974–6. *British Medical Journal*, **ii**, 1591–3.

Bartlett, J., Bridges, P., and Kelly, D. (1981). Contemporary indications for psychosurgery. *British Journal of Psychiatry*, **138**, 507–11.

Buckhardt, G. (1891). Uber Rindenexcisonen, als beitrag zur operativen therapie der psychosen. *Allgemeine Zeitschrift für Psychiatrie und psychischgerichtlithe Medicin*, **47**, 463–548.

Clare, A. J. (1980). Psychosurgery. In *Psychiatry in dissent*, pp. 278–342. Tavistock, London.

Eves, F. and Tata, P. (1989). Phasic cardiac and electrodermal reactions to idiographic stimuli in obsessional subjects. *Behavioural Psychotherapy*, **17**, 71–82.

Evans, B. M., Bridges, P. K., and Bartlett, J. P. (1981). Electroencephalographic changes as prognostic indicator after psychosurgery. *Journal of Neurology, Neurosurgery and Psychiatry*, **44**, 444–7.

Fulton, J. F. and Jacobson, C. G. (1935). The functions of the frontal lobes, a comparative study in monkeys, chimpanzees and man. *Advances in Modern Biology*, **4**, 113–23.

Hunter Brown, M. (1980). The captive patient: a forgotten man. In *The psychosurgery debate—scientific, legal and ethical perspectives* (ed. E. S. Valenstein), pp. 537–48. W. H. Freeman, San Francisco.

Hussain, E. S., Freeman, H., and Jones, R. A. C. (1988). A cohort study of psychosurgery cases from a defined population. *Journal of Neurology, Neurosurgery and Psychiatry*, **51**, 345–52.

Kelly, D. (1980). *Anxiety and emotions; physiological basis and treatment*. Thomas, Springfield; Ill.

Knight, G. (1965). Stereotactic tractotomy in surgical treatment of mental illness. *Journal of Neurology, Neurosurgery and Psychiatry*, **28**, 304–10.

Kringlen, E. (1970). Obsessional neurotics: a long-term follow-up. *British Journal of Psychiatry*, **111**, 709–22.

Marks, I. (1981). *Cure and care of neuroses; theory and practice of behavioural psychotherapy*, Wiley, New York.

Mindus, P., Bergstrom, K., Levander, S., Noren, G., Nyman, H., Persson, H. and Thomas, K-A., (1985). Clinical results and magnetic resonance findings in anxiety and obsessive-compulsive disorders treated with psychosurgery. In *Biological psychiatry—developments in psychiatry Vol. 7* (ed. C. Shagass, I. R. C. Josiassen, W. H. Bridger, K. J. Weiss, D. Staff, and G. M. Simpson), pp. 1370–2. Elsevier, New York.

Mitchell-Heggs, N., Kelly, D., and Richardson, A. (1976). Stereotactic limbic leucotomy—a follow-up at 16 months. *British Journal of Psychiatry*, **128**, 226–40.

Rachman, S. J. and Hodgson, R. J. (1980). *Obsessions and compulsions*. Prentice-Hall, Englewood Cliffs, NJ.

Richardson, A. (1973). Stereotactic limbic leucotomy: surgical technique. *Postgraduate Medical Journal*, **49**, 860–5.

Snaith, R. P., Price, D. J. E. and Wright, J. F. (1984). Psychiatrists' attitudes to psychosurgery. Proposals for the organisation of a psychosurgical service in Yorkshire. *British Journal of Psychiatry*, **144**, 293–7.

Tan. E., Marks, I., and Marset, P. (1971). Bimedial leucotomy in obsessive-compulsive neurosis, a controlled serial enquiry. *British Journal of Psychiatry*, **118**, 155–64.

Yaryura-Tobias, J. A. and Neziroglu, F. A. (1983). Psychosurgery. In *Obsessive-compulsive disorders. Pathogenesis, diagnosis and treatment* (ed. J. A. Yaryura-Tobias and F. A. Neziroglu), pp. 195–208. Marcel Deker, New York.

19

Chronic fatigue syndrome: can the psychiatrist help?

MICHAEL SHARPE

Introduction

Chronic fatigue syndrome (CFS) (Holmes *et al.* 1988) has been used as a diagnostic label for patients who have suffered from chronic (usually more than six months) disabling fatigue, for which medical assessment and investigation can find no adequate physical explanation. The symptom of fatigue can be applied to both mental and physical functioning (Wessely and Powell 1989). Patients with CFS have symptoms of mental fatigue such as difficulty in concentration and in thinking clearly, and of physical fatigue such as feeling weak. The physical fatigue in CFS appears not to be due to impaired muscle function and hence is 'central' in origin (Lloyd *et al.* 1988; Stokes *et al.* 1988). Other symptoms frequently reported include sore throat, headache, muscle pain, subjective fever, and swelling of joints and lymph glands.

The term CFS is preferable to those that embody unproven aetiological assumptions, such as 'post-viral fatigue syndrome', 'myalgic encephalomyelitis (ME)', 'post-infectious neuromyasthenia', and chronic Epstein–Barr syndrome'. As there are no reliably found abnormalities on physical examination or laboratory investigation, the syndrome is defined purely on the basis of symptoms (Wessely 1989).

There is evidence of an increasing number of people seeking medical help for chronic fatigue (Wessely 1989) and clinical experience indicates that patients with a main complaint of chronic disabling physical and mental fatigue are being referred in increasing numbers to psychiatrists, either by GPs or by hospital physicians. Such referrals are often made with little explanation of why a psychiatric opinion is requested, other than a negative medical assessment. Indeed these referrals may appear to reflect exasperation on the part of the referring doctor, rather than his or her faith in psychiatric treatment. Writing about loss of energy as a present complaint, a leading article stated: 'The physician's reaction is frequently one of frustration and helplessness, since he knows the bewildering variety of causes, the many psychological factors, and the frequent impotence of medical treatment' (Havard 1985).

The dilemma for the psychiatrist is how best to react to such referrals.

Should one also admit impotence, and perhaps refer the patient on to an interested physician, or should one accept the referral for treatment and risk frustration and helplessness when one finds oneself unable to alleviate the patient's symptoms.

In order to try to offer an answer to this dilemma, it is necessary to consider the following questions:

1. What is chronic fatigue syndrome?
2. What treatments are effective?
3. What is the best pragmatic management?

What is the chronic fatigue syndrome?

Fatigue is the central symptom of CFS. This is defined by the Oxford English Dictionary as 'weariness after exertion'. In practice, the symptom of fatigue has not been well defined and also refers to lack of energy, tiredness, and lassitude. Fatigue as a symptom is common: one survey found that 20 per cent of females and 14 per cent of males in a random sample of Americans aged between 25 and 74 complained of being 'tired or worn out' at least 'quite a bit of the time' (Chen 1986).

As well as being a common complaint, chronic fatigue is a recognized symptom of both physical and psychiatric illness. There are many physical illnesses and treatments that can cause fatigue (Plum 1982). Fatigue may also be a symptom of psychiatric disorder in the absence of physical disease. The symptom index of DSM-III-R (American Psychiatric Association 1987) lists 21 diagnostic categories associated with fatigue, the most common of which is depression.

The published research into CFS does not help greatly in clarifying its nature or cause and is difficult to interpret because it contains serious methodological flaws (David *et al.* 1988; Wessely 1989). The principal shortcomings have been: (a) poor case definition; (b) the use of non-random samples biased by being hospital referrals and often selected on the basis of exposure to a putative aetiological agent; (c) the use of inadequate control groups, leading to uncertainty about the specificity of any abnormalities found. Consequently there is no consensus about the causes and nature of CFS (Schwartz 1988). Thus if CFS as a syndrome is defined only on the basis of fatigue, combined with other 'non-specific' symptoms, could not such patients be suffering from well-recognized physical or psychiatric disorders, the appropriate treatment of which would relieve their symptoms?

The available evidence suggests that treatable physical disorders are found only in a small minority of cases (Taerk *et al.* 1987), whilst psychiatric diagnoses can be made in a high proportion (Bass 1989; White 1989). An initial attempt to produce an operational definition of CFS excluded patients with psychiatric diagnoses (Holmes *et al*. 1988). However the authors later acknowledged that physical and psychiatric diagnoses can coexist and recommended modification of the definition to permit their inclusion (Komaroff *et al*. 1989). The results of several studies suggest that a diagnosis of major depression can frequently be made (Taerk *et al*. 1987; Manu *et al*. 1988). Furthermore it occurs more often than in controls with peripheral neuromuscular disease even when fatigue is excluded as a diagnostic symptom (Wessely and Powell 1989).

Thus patients presenting with CFS may be suffering from other physical or psychiatric disorders (or both). However, despite the long list of potentially relevant physical diseases these are found in only a few. The important and potentially treatable disorders include anaemia, asthma, polymyalgia rheumatica, cancer, and myasthenia gravis. The majority of patients with CFS can be given psychiatric diagnoses, but it remains uncertain whether the psychiatric disorder is the primary cause of the symptoms or secondary to an as yet unidentified biological dysfunction. But management decisions cannot await the answer to this question. Many patients suffer serious disability (Behan *et al*. 1985), and there is evidence suggesting the prognosis is poor (Kroenke *et al*. 1988). Pragmatic management is needed now (White 1989).

What treatments are effective?

There have been few systematic evaluations of treatment in CFS. Those that have been tried can be seen to follow from underlying aetiological hypotheses. In order to clarify the treatment dilemma we are trying to resolve, treatments and hypotheses will be somewhat artificially divided into 'Medical' and 'Psychiatric'.

Medical treatment

A popular hypothesis has considered viral infection and associated immunological dysfunction as central to the disorder (Behan *et al*. 1985). Medical treatments have assumed that such infection not only precipitated the illness, but also continues to maintain it. Hence eradication of the putative virus should lead to recovery. The systematic studies of such treatments have however been disappointing. Al Kadiry *et al*. (1983) reported negative results from a trial of immunoglobulin. A recent, well-designed trial of Acyclovir also failed to show any benefit over placebo, although many patients showed a transient improvement with both treatments (Straus *et al*. 1988). Several patients receiving Acyclovir suffered reversible renal failure. So far, despite enthusiasm for a number of medical treatments there is a lack of convincing evidence for their efficacy, other than as placebos, and they may be hazardous.

Psychiatric treatment

As the aetiology of most psychiatric disorders is little understood this form of treatment tends to be pragmatic. Applied to CFS the assumption is that whatever the precipitant of the syndrome, the symptoms and disability may be perpetuated by associated psychiatric disorder, the patients beliefs, their behaviour, and their relationships. Treatments aimed at modifying these factors may be divided into pharmacological and psychological.

Pharmacological treatments   Antidepressant drugs have been tried in patients with CFS, and a claim of a 70 per cent response rate to low dose amitriptyline has been made (Jones and Straus 1987, pp. 102–4), but the supporting evidence is unpublished. There have been no controlled trials of antidepressants in CFS. There are however two published controlled trials of amitriptyline in patients with the 'fibromyalgia syndrome'. This is a clinically defined disorder characterized by fatigue, aches, tenderness, and disturbed sleep (Yunus 1989) and potentially indistinguishable from CFS (Pritchard 1988). Both these trials found amitriptyline in low dose (25–50 mg per day) to be superior to placebo (Goldenberg *et al.* 1986; Carette *et al.* 1986).

The monoamine oxidase inhibiting antidepressant drugs such as phenel-zine are currently receiving renewed attention (Bass and Kerwin 1989). A priori their energizing effect may be considered useful in CFS, but again there is no systematic evidence to support their use. It is worth noting however that phenelzine was found to be superior to imipramine in patients with atypical depression, defined as RDC depressive disorder with atypical symptoms including severe fatigue (Liebowitz *et al.* 1988). It is possible therefore that there is a role for phenelzine in some patients with CFS.

Individual psychological treatment   Here again there are no controlled trials to guide us, but a recently described form of treatment is of interest (Wessely *et al.* 1989). It is based on a cognitive-behavioural approach to the maintenance of symptoms, but also includes supportive psychotherapy, involvement of the spouse, and drug treatment of depression.

The cognitive-behavioural approach is based on characteristics observed in a sample of patients seen at the National Hospital for Nervous Diseases (Wessely and Powell 1989; Wessely *et al.* 1989). These are first, the avoidance of exercise; and second, the beliefs held by patients that (a) exercise is harmful, (b) symptoms such as tachycardia, and muscle ache arising during or after exercise indicate relapse, and (c) symptoms are caused by a biological process outside their control, and best combated by medical (or 'alternative medical') treatment.

It is hypothesized that the lack of activity is a cause of fatigue and leads to symptoms when exercise is attempted because of lack of fitness. The reduced activity will also lead to an absence of reward and contribute to low mood which in turn will contribute to fatigue. Hence lack of exercise, fatigue, and low mood are interrelated in a vicious circle.

From the above it follows that the aim of treatment is to help patients to increase their level of activity and to reappraise beliefs that impede this change. To this end patients are offered two alternative explanations for their symptoms.

1. The conventional view: they are suffering from a poorly understood disease. Physical and mental activity should be limited and they should respond to any increase in symptoms by further rest.

2. The alternative view: they have probably had an acute illness (perhaps an infection) which forced them to become inactive for a period of time. They have become unfit and consequently experience symptoms when attempting activities, but never pursue them long enough to allow the symptoms to subside. These symptoms are real, but incorrectly attributed to a recurrence of the original infection. Hence they are caught in a cycle of increasing avoidance of exercise, inactivity, fatigue, and perhaps depression.

Patients are then encouraged to decide which of these is the most helpful explanation of their symptoms by collaborating with the psychiatrist in a behavioural experiment. The experiment is gradually and consistently to increase their level and range of physical and mental activity, and to monitor their performance. Cognitive psychotherapy continues in parallel, to help the patients to challenge the attribution of their symptoms to relapse, to re-evaluate their beliefs about the nature of the illness in the light of their improvement, and to encourage them to persist in the exercise programme.

To date about 50 patients have been treated using this approach. The unpublished results (Wessely, personal communication) indicate that a third of patients refused to engage in treatment. Of those that did engage, two-thirds have shown a substantial and persisting improvement in disability, some returning to work. No patient has deteriorated.

There are some caveats to be made about this treatment approach. The sample of patients was acquired from those referred to the National Hospital for Nervous Diseases, and therefore is likely to be atypical. A substantial proportion of patients refused treatment. Furthermore, the treatment has not yet been evaluated in a controlled fashion, and it is unclear which ingredients of the 'package' are most important. However, although this treatment clearly requires further evaluation the initial reports are sufficiently encouraging to support its clinical application on an empirical basis.

Marital and family therapy There is no published evidence on which to evaluate the specific role of these treatments in patients with CFS. As in any chronic disabling condition there is inevitably a major impact on other family members, and changed demands on, and expectations of the patient. Counselling and education of the family are likely to be beneficial both to the patient and other family members. It is certainly preferable for the spouse or partner to be closely involved in treatment. There is evidence for the efficacy of this aspect of treatment in patients disabled as a consequence of stroke (Evans *et al.* 1988).

Clinical guidelines

The evaluation of treatments for CFS is only just beginning, but the evidence available suggests that both antidepressant drugs and a cognitive-behavioural psychotherapeutic approach are useful. In deciding whether to offer treatment the psychiatrist should bear in mind the often severe disability associated with CFS, the poor prognosis, the lack of an effective medical treatment, the potential morbidity from experimental medical treatments, and preliminary reports of effective psychiatric treatment.

I suggest that in the light of these points it is reasonable for a psychiatrist to offer treatment.

The following are guidelines for management.

1. *Exclude treatable physical illness*

This is largely the responsibility of the referring physician, but the psychiatrist should ensure that it has been done. Physicians who consider any assessment pointless because of the low yield of such screening should be specifically requested to make an adequate assessment before psychiatric treatment is commenced. More often patients will have received extensive investigation before coming to a psychiatrist and it may be necessary to persuade the patient to desist from further pursuit of the 'definitive test'. The psychiatrist would be well advised to keep an open mind about physical disorder, but to initiate or sanction further medical assessment only if there are clear indications, and not simply to reassure the patient.

2. *Engage the patient in treatment*

This is crucial. It can be facilitated if the referral is seen by the patient to be made, and greeted, with optimism and enthusiasm. Clinical experience suggests that a fear of psychiatric illness is especially prominent in this group of patients. A joint consultation with physician and psychiatrist is a useful way of overcoming patients' reluctance to be referred. An argument at this point over whether symptoms have a physical or psychological cause can be disastrous for the collaborative relationship necessary for successful treatment. It is important to acknowledge the reality of the patient's symptoms, whilst advocating a pragmatic approach to treatment.

3. *Treat identified psychiatric syndromes*

If the patient has symptoms of a depressive disorder a trial of an anti-depressant drug is worthwhile. A tricyclic antidepressant may be tried in the first instance, the evidence presented above suggesting that a low dose may suffice in some cases. An MAOI such as phenelzine may be considered if the features of atypical depression (p. 18) are present. Intolerance of side-effects may be a problem. Patients can be helped to tolerate antidepressants by careful education and discussion aimed at preventing side-effects being incorrectly interpreted as representing a deterioration in the illness. In addition drugs that are less sedative and with fewer physical side-effects (such as lofepramine) are more easily tolerated. Anxiety may also be treated by antidepressant drugs (p. 43), but if prominent may respond to a psychological approach such as anxiety management training (Clark 1989, p. 52).

4. *Treat exercise avoidance and associated beliefs*

A graded increase in the patient's level of activity is an essential component of management. This exercise programme should be designed to start from the patient's current activity level, however low. Treatment for depression can continue concurrently, although if severe the depression may be best treated first. Exercise targets should be agreed by patient and psychiatrist, and it is preferable to choose activities the patient will find rewarding. Exercise should be daily and progress by small and graded increments. Gains should be maintained for at least a week before the targets are increased. The patient should be asked to keep records of his or her activities, and also consequent symptoms. These records can then be reviewed by psychiatrist and patient together, every one to two weeks. Symptoms arising from exercise should be reviewed using the cognitive psychotherapeutic approach described. Treatment can be expected to require at least six one-hour sessions.

Tackle marital and family maintaining factors

Clinical experience strongly suggests that these factors may act so as to maintain symptoms, and to undermine treatment. Family members should be involved in psychological treatment where appropriate. The model of treatment should be explained to them and they may then act as 'co-therapists'. In some cases the family may be observed to be acting so as to oppose the patient taking a more active role. In such cases family or marital therapy may be indicated. The task of this therapy is to elucidate how the family may be unwittingly retarding the patient's recovery, and to facilitate change in the family system, so that pressures opposing the patient's rehabilitation are reduced. (For further details of one approach to family therapy, see Procter and Walker (1988)).

238 *Dilemmas in the management of psychiatric patients*

6. Tackle social maintaining factors

'ME' has become something of a fashionable disorder in recent years. In the UK there are two patients' support groups: the ME Association and the ME Action Campaign. To date they have advised sufferers to rest and have tended to be rather disparaging of psychiatric treatment—although this situation may be changing, at least with regard to the use of anti-depressant drugs. None the less patients may need to be encouraged to be as sceptical of the advice of these groups and the alternative therapies they recommend, as they are about 'psychiatric treatments'.

7. Beware of pitfalls in management

The main pitfall is probably failure to engage the patient in treatment. Even when apparently engaged a proportion will drop out of treatment. Taking the patient's symptoms and their understanding of the disease seriously, whilst offering the alternative viewpoint (p. 235). will help to minimize this attrition.

Patients may be reluctant to take antidepressant medication. Such reluctance may be countered by emphasizing the understandability of depression as a consequence of their disability.

Setting appropriate goals for exercise is important. Some patients may find it difficult to increase their level of exercise, while others will try to make rapid, large increases, with the risk of subsequent 'relapse'. The psychiatrist can help both groups of patients avoid demoralization by encouraging small increases, but on a daily basis. If this is not done a see-saw pattern may result with patients overreaching themselves on one day and spending the next in bed.

The psychiatrist should also beware of the emergence of physical disorder during treatment, as even after extensive pre-treatment investigation unsuspected medical conditions may occasionally emerge.

The clinical guidelines above will need to be updated as new evidence emerges. However it is hoped that even in their current form they will offer a strategy of management to psychiatrists treating patients with CFS, and allow them to do so without risk of being either paralysed by therapeutic impotence or frustrated by therapeutic failure.

References

Al Kadiry, W., Gold, R. G., Behan, P. O., and Mowbray, J. F. (1983). Analysis of antigens in the circulating immune complexes of patients with Coxsackie Infections. In *Immunology of nervous system infections. Progress in brain research 59* (ed. P. O. Behan, V. ter Meulen, and F. C. Rose), pp. 61–7. Elsevier Science Publishers, Holland.

American Psychiatric Association (1987). *Diagnostic and statistical manual of mental disorders.* Third edition, revised. American Psychiatric Association, Washington DC.

Bass, C. (1989). Fatigue states. *British Journal of Hospital Medicine*, **41**, 315.

Bass, C. and Kerwin, R. (1989). Rediscovering monoamine oxidase inhibitors. *British Medical Journal*, **298**, 345–6.

Behan, P., Behan, W., and Bell, E. (1985). The post-viral fatigue syndrome—An analysis of the findings in 50 cases. *Journal of Infection*, **10**, 211–22.

Carette, S., McCain, G. A., Bell, D. A., and Fam, A. G. (1986). Evaluation of amitriptyline in primary fibrosis. *Arthritis and Rheumatism*, **29**, 655–9.

Chen, M. K. (1986). The epidemiology of self-perceived fatigue among adults. *Preventive Medicine*, **15**, 74–81.

Clark, D. M. (1989). Anxiety states: panic and generalized anxiety. In *Cognitive behaviour therapy for psychiatric problems: a practical guide* (eds. K. Hawton, P. Salkovskis, J. Kirk, and D. M. Clark), pp. 52–96. Oxford University Press.

David, A. S., Wessely, S. and Pelosi, A. J. (1988). Post-viral fatigue syndrome; time for a new approach. *British Medical Journal*, **296**, 696–8.

Evans, R. L., Matlock, A. L., Bishop, D. S., Stranachan, S., and Pederson, C. (1988). Family intervention after stroke: does counselling or education help? *Stroke*, **19**, 1243–9.

Goldenberg, D. L., Felson, D. T., and Dinerman, H. (1986). A randomized, controlled trial of amitriptyline and naproxen in the treatment of patients with fibromyalgia. *Arthritis and Rheumatism*, **29**, 1371–7.

Havard, C. W. H. (1985). Lassitude. *British Medical Journal*, **290**, 1161–2.

Holmes, G. P., *et al.* (1988). Chronic fatigue syndrome: a working case definition. *Annals of Internal Medicine*, **108**, 387–9.

Jones, J. F. and Straus, S. E. (1987). Chronic Epstein–Barr virus infection. *Annual Reviews in Medicine*, **38**, 195–209.

Komaroff, A. L., Straus, S. E., Gantz, N. M. and Jones, J. F. (1989). The chronic fatigue syndrome. *Annals of Internal Medicine*, **110**, 407–8.

Kroenke, K., Wood, D. R., Mangelsdorf, D., Meier, N. J., and Powell, J. B. (1988). Chronic fatigue in primary care. *Journal of the American Medical Association*, **260**, 929–34.

Liebowitz, M. R., *et al.* (1988). Antidepressant specificity in atypical depression. *Archives of General Psychiatry*, **45**, 129–37.

Lloyd, A. R., Phales, J., and Gandevia, S. C. (1988). Muscle strength, endurance and recovery in the post-infection fatigue syndrome. *Journal of Neurology, Neurosurgery and Psychiatry*, **51**, 1316–22.

Manu, P., Matthews, D. A., and Lane, T. J. (1988). The mental health of patients with a chief complaint of chronic fatigue. *Archives of Internal Medicine*, **148**, 2213–17.

Plum F. (1982). Asthenia, fatigue and weakness. In *Cecil's textbook of medicine, vol. 2* (ed. J. B. Wyngaarden and L. H. Smith (Jr)), pp. 1968–9. W. B. Saunders Co., Philadelphia.

Pritchard, H. (1988). Fibrositis and the chronic fatigue syndrome. *Annals of Internal Medicine*, **106**, 906.

Procter, H. and Walker, G. (1988). Brief therapy. In *Family therapy in Britain* (ed. E. Street and W. Dryden), pp. 127–49. Open University Press, Milton Keynes.

Schwartz, M. N. (1988). The chronic fatigue syndrome—one entity or many? *New England Journal of Medicine*, **319**, 1726–8.

Stokes, M. J., Cooper, R. G., and Edwards, R. H. T. (1988). Normal muscle strength and fatiguability in patients with effort syndromes. *British Medical Journal*, **297**, 1014–16.

Straus, S. E., *et al.* (1988). Acyclovir treatment of the chronic fatigue syndrome: lack of efficacy in a placebo controlled trial. *New England Journal of Medicine*, **319**, 1692–8.

Taerk, G. S., Toner, B. B., Salit, I. E., Garfinkel, P. E., and Ozersky, S. (1987). Depression in patients with neuromyasthenia (benign myalgic encephalomyelitis). *International Journal of Psychiatry in Medicine*, **17**, 49–56.

Wessely, S. (1989). Myalgic encephalomyelitis—a warning: discussion paper. *Journal of the Royal Society of Medicine*, **82**, 215–17.

Wessely, S. and Powell, R. (1989). Fatigue syndromes: a comparison of chronic 'post-viral' fatigue with neuromuscular and affective disorders. *Journal of Neurology, Neurosurgery and Psychiatry*, **52**, 940–8.

Wessely, S., David, A., Butler, S., and Chalder, T. (1989). Management of chronic (post-viral) fatigue syndrome. *Journal of the Royal College of General Practitioners*, **39**, 26–9.

White, P. (1989). Fatigue syndrome: neurasthenia revisited. *British Medical Journal*, **298**, 1199–200.

Yunus, M. B. (1989). Fibromyalgia syndrome: new research of an old malady. *British Medical Journal*, **298**, 74–5.

20

Confidential information: when should it be disclosed?

DAVID JULIER

The problem

Most members of the public hold the romantic belief that every doctor qualifying takes the Hippocratic Oath: 'Whatever, in connection with my professional practice, or not in connection with it, I see or hear, in the life of man, which ought not to be spoken of abroad, I will not divulge, as reckoning that all such should be kept secret.' But things have moved on since 420 BC. The closely circumscribed relationship between doctor and patient has become a complex interaction involving many diverse professionals handling widely diffused information and subjected to many social pressures and expectations. Ethical problems are correspondingly more complex.

The central difficulty can be stated thus: current codes of ethical practice in this country are united in the view that information about patients may be disclosed only in exceptional circumstances; there is, however, an absence of unanimity and often of adequate detail in the description of these exceptions; as a result clinicians frequently find themselves uncertain whether a particular clinical situation is covered by an accepted code of conduct or whether some individual course of action must be improvised; and if the latter then how taxing may be the General Medical Council's (GMC) stipulation that 'Whatever the circumstances, a doctor must always be prepared to justify his action' (General Medical Council 1987).

That there is in respect of confidentiality such an area of doubt should not be a source of dismay. The GMC (1988) accepts this: 'In all areas of medical practice doctors need to make judgements which they later have to justify. This is true both of clinical matters and of the complex ethical problems which arise regularly in the course of providing patient care, because it is not possible to set out a code of practice which provides solutions to every such problem which may arise'. Furthermore in the UK, ethical systems are not codified within the civil and criminal law but depend on professional guidance channelled through the GMC which, under the Medical Act of 1858, is responsible to the Privy Council for enforcing professional standards. This arrangement promotes flexibility and responsiveness to changing conditions, but sometimes at the expense of clarity and certainty.

What will be attempted in this chapter is, firstly, to set out what may be stated with some certainty about the circumstances in which it is permissible to disclose confidential information, and, secondly, to discuss some common situations where uncertainty is more evident.

Useful publications

Guidance is to be found in the GMC's booklet *Professional conduct and discipline: fitness to practise* (GMC 1987), in the British Medical Association's *Philosophy and practice of medical ethics* (BMA 1988), and in pamphlets from the medical insurance societies (one of these organizations should of course be consulted when there is doubt about a practical situation). The most detailed guidance is to be found in the Report from the Confidentiality Working Group produced by the Steering Group on Health Service Information (Körner 1984); copies of this lucid document are in regional health authority libraries.

Two important codes will be published soon: the Interprofessional Working Group's *Code on confidentiality of personal health information*, and the code concerning patients' right of access to their personal medical records.

Circumstances in which confidential information may be disclosed

To address our first task, it seems that most guidelines are in agreement in suggesting that information may be disclosed in the following circumstances (although their grouping under the ten headings used here is not part of an agreed convention).

The patient gives valid consent

This simple statement encompasses many issues which demand scrupulous attention. Clearly your patient needs to understand the nature, purpose, and implication of the disclosure, what factors or opinions (at least in outline) you propose to convey, who is the recipient and what use that person will make of the communication. For your part, you need to satisfy yourself that disclosure is appropriate and that your patient's consent is not given in a spirit of passive compliance, under the influence of medication or under duress, for example, where he or she perceives consent to be a condition of continued treatment.

The relatives' role   On all these issues there is scope for uncertainty. Can this be circumvented by securing the nearest relative's agreement? Clearly a relative may make the issues more understandable for the patient (and indeed for you), but equally clearly the relative cannot give consent on behalf of your patient. If valid consent cannot be obtained, the question is whether you decide yourself on disclosure in your patient's interest; this is discussed later.

Access to Medical Records Act 1988   This entitles a patient to read a report prepared (with consent) for insurance or employment purposes and to withdraw consent for its submission or request amendment of inaccurate statements. You may withhold information from the patient if, as a matter of clinical judgement, the disclosure would be likely to cause serious harm to the individual's physical or mental health.

The patient's consent may be assumed to be implicit

Three types of situation fall under this heading. Health professionals and other staff involved in the patient's care will need to share information. The GMC (1987) specifies: 'To the extent that the doctor deems it necessary for the performance of their particular duties, confidential information may be shared with other professionals . . . who are assisting and collaborating with the doctor in his professional relationship with the patient.' The second situation is where health service professionals and students need access to patient data for educational purposes. The third concerns social workers involved with patients in health service settings. These topics will be considered in more detail later.

Disclosure is in the interests of the patient or of the general public

This disconcertingly broad heading gives plentiful scope for uncertainty and the exercise of judgement. Let us consider, for example, whether we can assume it is in our patient's interest to have us inform his or her relatives of the nature of the patient's illness. The Royal College of Psychiatrists (1985) is encouraging: '. . . the balance of emphasis should be on giving as much information as possible. Withholding it should be the exception; and founded on a clear likelihood that divulgence would do more harm than good. Clearly, families are not on a par with outsiders'. Certainly we can agree that families are capable of understanding and support, but equally they are capable of emotional over-involvement, aggressive criticism, scapegoating, and covert conspiracy, so that our patient may have strong views about what we tell the devoted relatives. Convention dispenses with the patient's explicit consent but this must be replaced by sensitivity to the individual circumstances of each case.

On the other hand, issues of confidentiality within a family or other social group are clearer when the health or safety of a child is at risk. The Royal College of Psychiatrists (1988) strongly supports disclosure: 'The welfare and best interests of the child have to be regarded as of the first importance, superseding duties of confidentiality which the doctor may have to other persons, including parents.' We should of course seek to retain our sympathy towards those whose confidence we have to betray.

Public interest   Where disclosure is made in the public interest it has to be sustainable that the risk to the public is so serious as to prevail over the patient's right to confidentiality. The assessment of this risk may be no easy matter, particularly in the context of psychotherapy where the exercise of fantasy is given some licence. Disclosure is likely to terminate the doctor–patient relationship.

Three types of issue will serve as examples. Firstly, in relation to the prevention, detection, and prosecution of crime you need to be convinced that the actual or potential crime is of an extremely serious nature, but how do you judge that? Section 116 of the Police and Criminal Evidence Act 1984 gives some guidance on what constitutes a 'serious arrestable offence' but also leaves much scope for doubt. Secondly, where a patient ignores advice that he or she is unfit to drive because of a medical condition, you may be justified in submitting a report to the Driver and Vehicle Licensing Centre without consent. Thirdly, disclosure may be justified in the prevention and control of communicable diseases.

The information is to be used for medical research

Confidentiality is safeguarded by the requirement that every research project should be approved by a local ethical committee. This committee must satisfy itself that patients give consent or that there are good grounds for dispensing with consent, that data are stored safely and destroyed at the end of the project, and that no patient could be identified from published material.

Under the subject access provisions of the Data Protection Act 1984

An individual is entitled to know whether a health authority holds personal information on him or her in a computer system, and may claim right of access to that information. However the Data Protection Subject Access Modification Health Order 1987 permits the health authority, on advice given by a senior health professional, to omit from disclosure personal data which 'would be likely to cause serious harm to the physical or mental health of the data subject'.

Disclosure of patient data is required by statute

This includes evidence to Courts about offenders thought to be mentally disordered (covered by Sections 35, 36, 37, 38, and 41 of the Mental Health Act 1983); the requirement to report to the parents and the local education authority the opinion that a child under the age of five has or probably has special educational needs; and the obligation to disclose information which would be of material assistance in preventing terrorism or in securing the arrest of a person involved in an act of terrorism.

Under this heading we are concerned not with the authorization but with the legal requirement to disclose information, and this applies equally to the notification of drug addicts. Under the Misuse of Drugs Act 1971 and Misuse of Drugs Regulations 1973 we have by statute to notify to the Home Office patients known or suspected to be addicted to certain controlled drugs; principally opiates, certain opiate derivatives, and cocaine.

Notification must be repeated at 12-monthly intervals if you remain in contact with the patient. This Home Office information carries no legal consequences for the addict, but it can be accessed by doctors able to satisfy the Home Office that it is required for the patient's clinical management.

Disclosure is required by bodies with statutory powers

These include Coroners' Courts and other Courts of Law where you may well find yourself facing the Court's expectation that you should reveal confidential material acquired in the context of a therapeutic relationship, when you may consider such disclosure unethical. As a doctor and psychiatrist you enjoy no legal privilege, but the Court may listen sympathetically to your request to withhold certain information. If the Court should decline your request the question is whether it would then seem right to opt for disclosure as being legal but ethically wrong, or silence which is ethically correct but legally in contempt of Court. Happily such a dilemma arises only rarely.

Statutory powers can also be exercised by the Health Service Commissioner, an Inquiry appointed by the Secretary of State for Social Services, the Health and Safety Commission and Executive, and Employment Medical Advisers (this relates to the school medical record of a person under 18), and professional bodies of the various health professions which may issue writs of subpoena in relation to disciplinary proceedings, and these bodies could require medical records to be produced.

Disclosure is required by bodies with statutory access

This allows the Mental Health Act Commission and Mental Health Review Tribunals to inspect psychiatric records. There are also situations in which the patient's representative may ask to see the records, for example, when obtaining an independent medical report on the patient's condition. Medical and social reports to the Tribunal must also be copied to the applicant and the patient, or to his legal representative. However, you are permitted to suggest to the Tribunal that certain parts of the report should be withheld from the patient where the information would adversely affect the patient's health or welfare.

Access is required by health authority managers for essential management functions

These would include investigating a complaint or untoward incident, responding to a claim for compensation, and exercising various powers under the Mental Health Act, for example an appeal to the Hospital Managers.

Disclosure is required in order to safeguard national security

The scenario is a rare one: questions are raised about the fitness of the most senior politician in the land to remain in office; he or she is your patient; you receive a certificate signed by a Cabinet Minister or the Attorney General or the Lord Advocate stating that disclosure of your psychiatric records is required to safeguard national security. Do not hesitate: you are about to achieve fame (or of course disgrace).

Some areas of particular difficulty

In this section we shall consider a number of clinical situations in which there is considerable scope for uncertainty as to how we should apply the principles outlined so far.

Should we record sensitive information in the medical notes?

Let us imagine that in the course of your follow-up of a patient with a manic-depressive disorder she indicates her wish to talk in detail about her experience of being abused as a child by her father, and to have this information treated as very confidential. You reflect that to record the information in her medical notes amounts to a form of disclosure, albeit limited to professional staff bound by a code of ethics similar to your own. You are of course under no formal obligation to write notes; but to maintain no record of consultations might be considered negligent in the event of medico-legal proceedings, and in any case this practice would make heavy demands on memory. You cannot offer yourself the option of destroying the medical record once the patient is discharged, for the record is the property of the Secretary of State (this ownership function being devolved to the health authority and thence to the unit records officer). There is, perhaps, no satisfactory solution: the least unsatisfactory solution—and one used commonly in the practice of psychotherapy—may be to note in the medical record 'Detailed background information with Dr X', and to keep a separate note of the details in a locked cabinet, access to which is controlled by the responsible consultant (Royal College of Psychiatrists, 1977).

Confidentiality within a multidisciplinary team

Let us imagine now that you see this patient in a community setting where you operate a multidisciplinary team approach; team meetings provide for the supervision of all case work, and all team members, including the psychiatrist, would normally report their cases in detail so that treatment plans are discussed and agreed.

Your dilemma is this: if you withhold the sensitive information as you present your patient's case to other team members you may invalidate their contribution to the treatment plan, and—perhaps more disconcertingly—establish a precedent whereby other team members keep back information in a way which could be dangerous (for example, with a case involving current child abuse); on the other hand to share the information which your patient sees as very private might seem at odds with her expectation of a confidential relationship with a doctor, and entails the risk that team members may in turn divulge the information to others.

Of course it may have been possible to anticipate the dilemma by telling your patient in advance about the usual practice within the team, but there could be occasions when you would argue that this is distracting and confusing for a patient. In the absence of such an explicit discussion would you be justified in assuming her implicit consent? The argument would rest on viewing team members as 'professionals ... who are assisting ... the doctor in his professional relationship with the patient'. Their task is indeed to help devise the most effective treatment, so that the principle of implied consent seems to be applicable.

Medical responsibility   Now what if there should be subsequently a breach of confidentiality in that a team member divulges the information to someone not associated with the care of the patient—are you to be held responsible? This might hinge on two issues. The first is embodied in the GMC (1987) code: 'It is the doctor's responsibility to ensure that such individuals (that is, other health professionals) appreciate that the information is being imparted in professional confidence.' This would seem to require both that the multidisciplinary team has a continuing understanding and agreement on confidentiality in general, and that specific items of information are highlighted as being particularly sensitive.

Shared responsibility   The second issue concerns the membership of the team. Most members of other health care professions implicitly acknowledge codes of practice, and these are specified by the General Nursing Council and the Council for Professions Supplementary to Medicine (the latter include chiropodists, dieticians, scientific officers in laboratories, occupational therapists, physiotherapists, and remedial gymnasts). For psychologists, membership of the British Psychological Society has implied acceptance of a code, but in the future this function may be covered more comprehensively by the register of chartered psychologists. In addition any professional holding a contract with a health authority is expected to maintain confidentiality as a condition of employment, and increasingly the trend is for this to be made explicit in contracts. Social workers are of course employed by the local authority but are similarly bound by their conditions of employment, and for some by the code of practice of the British Association of Social Workers.

All these contracted professional members of a multidisciplinary team must be seen as maintaining their own responsibility for the confidentiality of information, however acquired. The psychiatrist's responsibility is limited to the initial decision to disclose to the team, and to his or her long-term role in contributing to the monitoring and maintenance of ethical and clinical standards in the team; this latter role must entail keeping ethical issues on the agenda of the team's discussions and analysing any apparent breaches of confidentiality. The position in relation to students and volunteers attached to the team is a little different in that their behaviour may in some cases not be linked to contractual or professional sanctions. This means that a greater level of continuing responsibility must rest with the professional who discloses information to the student or volunteer, and with the professional supervising that individual.

The multidisciplinary case conference

You are invited by the local authority social services department to participate in a case conference concerning your manic-depressive patient: her children have been taken into care because it is thought she has neglected them at home. You note that participants at the meeting include your patient's home-help and her children's school counsellor and foster-mother.

Clearly it is important for you to attend, and clearly you would hope to secure her agreement to your going and disclosing at least some basic medical information about her illness; but should you go if she withholds her consent? You could of course go as a silent observer, but this might condemn you to watch helplessly as some disastrously inappropriate decision was reached. We have already noted that confidential information may be shared with other professionals who are assisting and collaborating with the doctor, and you may argue that this applies despite the fact that the meeting is convened by another agency. Alternatively you may rely on the guidance that under exceptional circumstances disclosure may be made without consent in the best interests of the patient. This could apply if it was likely, for example, that your information would lead to the children being returned to their mother, which is what she would wish and would otherwise not achieve. More usually you would have in mind that disclosure may be justified 'in the public interest which . . . might override the doctor's duty to maintain his patient's confidence' (GMC 1987). The public interest here would be the children's safety and well-being, and this issue will be reassuringly clear-cut (albeit potentially distressing to the patient) when there is a risk of child abuse.

Non-professional participants You can thus find grounds for sharing at least certain key pieces of information with other health care professionals and social workers, but you may well have reservations in principle about other participants at the case conference, since these may not be bound by an acknowledged code of practice. Your solution to this difficulty may be to contact the convenor in advance to discuss how you might present your medical evidence before the arrival or after the departure of the foster-mother, the home-help, and possibly the school counsellor.

Confidential minutes   At the close of the case conference you will need to enquire of the convenor who is to receive copies of the Minutes, making your point that the distribution should be limited to health care professionals and social workers. You should also check that the Minutes will be marked 'Confidential' so that they would not be produced automatically for your patient to read should she demand right of access to her Social Services Department records (in accordance with the circular: Personal Social Services Records—Disclosure of Information to Clients, HC(83)20/LAC(83)14), although in most cases it would be proper for her to know what was said. Access to these records may also be claimed by a local authority councillor (this right of access was upheld by a Birmingham court in 1983), but local codes of practice would usually make an exception for confidential medical reports.

HIV infection and AIDS

Envisage the situation in which a patient admitted with anxiety symptoms associated with a fear of contracting AIDS is found on investigation (carried out with his consent) to be HIV positive. He refuses to allow you to tell his wife and general practitioner. Should you none the less inform them?

Disclosure to primary care team   Let us explore first the question as it relates to the primary care team. From the legal point of view disclosure is permitted. The National Health Service (Venereal Diseases) Regulations 1974 specify that health authorities are required to ensure that information about persons examined or treated for sexually transmitted diseases should be treated as confidential. Disclosure is, however, permitted to a doctor or person employed under the direction of a doctor in connection with and for the purpose of treatment or prevention of the spread of the disease. However, the GMC (1988) advocates a cautious approach: '. . . the patient should be counselled about the difficulties which his or her condition is likely to pose for the team responsible for providing continuing health care and about the likely consequences for the standard of care which can be provided in the future. If, having considered the matter carefully . . . the patient still refused to allow the general practitioner to be informed then the patient's request for privacy should be respected.' Thus we are encouraged to accord a special level of privacy to those with HIV infection even though this must jeopardize the medical and psychological help they can get from others who remain ignorant of the diagnosis. At the same time the

GMC 'expects that doctors will extent to patients who are HIV positive or are suffering from AIDS the same high standard of medical care and support which they would offer to any other patient'.

The GMC advises, however, that an exception to secrecy arises 'where the doctor judges that the failure to disclose would put the health of any of the health care team at serious risk . . . in such a situation it would not be improper to disclose such information as that person needs to know'. Now in order to satisfy ourselves that there is not a serious risk to the primary care team we have to assume either that our patient will appropriately and unfailingly advise any team member who proposes to undertake some procedure which could pose a risk (and his refusal of consent of disclosure would cast some doubt on this), or that the team would in every circumstance and with every patient take those precautions which would be relevant with a patient known to be HIV positive. The latter issue could only be clearly resolved by being put to the general practitioner (if this can be done whilst preserving the patient's anonymity). It would then surely be a brave doctor who confidently replied that ignorance would not put his or team at risk.

It therefore seems likely that in almost every case disclosure may be justifiable, and if justifiable then probably ethically obligatory.

Disclosure to sexual partner   Similar arguments may be adduced in favour of informing our patient's wife. The GMC's view is that 'there are grounds for such disclosure only where there is a serious and identifiable risk to a specific individual who, if not so informed, would be exposed to infection . . . where such consent is withheld the doctor may consider it a duty to seek to ensure that any sexual partner is informed, in order to safeguard such persons from a possibly fatal infection'. This view is endorsed by Black (1989).

On the other hand you may argue that by informing your patient's wife you would be doing more than protect her against infection. This may represent a major intervention in the marital relationship with far-reaching emotional and other consequences. Moreover to take upon yourself a role which would normally rest with your patient may be to deprive him of a proper sense of responsibility for his sexual behaviour, and it may be vital that he faces that responsibility himself since he may otherwise be a risk to other sexual partners whom you could not warn.

Patient with incapacity   The question may of course present differently if your patient has some cognitive impairment or a persisting functional illness making it impossible for him to grasp the issues. You may then need to set yourself the task of taking that decision (if it cannot be deferred) which you believe he might have taken himself when lucid. Your reasoning is likely to be heavily influenced by your view of the ideal moral position.

Ideally it seems to me the wife should be informed, but maybe the doctor should take this upon himself only when the patient is incapable by virtue of mental disorder of taking and implementing this decision.

References

Black, D. (1989). Ethics of HIV testing. *Journal of the Royal College of Physicians of London*, **23**, 19–21.

British Medical Association (1988). *Philosophy and practice of medical ethics*. British Medical Association. London.

General Medical Council (1987). *Professional conduct and discipline: fitness to practise*. General Medical Council, London.

General Medical Council (1988). *HIV infection and AIDS: the ethical considerations*. General Medical Council, London.

Körner, E. (1984). *Protection and maintenance of confidentiality of patient and employee data: a report from the Confidentiality Working Group*. Department of Health and Social Security, London.

Royal College of Psychiatrists (1985). *A Guide to confidentiality in relation to mental health: discussion document from a Joint Working Party*. (Unpublished report).

Royal College of Psychiatrists (1988) Child psychiatric perspectives on the assessment and management of sexually mistreated children. Prepared by a working group of the Child and Adolescent Specialist Section. *Psychiatric Bulletin of the Royal College of Psychiatrists*, **12**, 534–540.

Index